FROM AFFINE TO EUCLIDEAN GEOMETRY

AN AXIOMATIC APPROACH

WANDA SZMIELEW

FROM AFFINE TO EUCLIDEAN GEOMETRY

AN AXIOMATIC APPROACH

Edited, prepared for publication and translated from the Polish by

MARIA MOSZYŃSKA
Warsaw University

D. REIDEL PUBLISHING COMPANY
DORDRECHT : HOLLAND / BOSTON : USA / LONDON : ENGLAND

PWN–POLISH SCIENTIFIC PUBLISHERS
WARSZAWA : POLAND

Library of Congress Cataloging in Publication Data

CIP

Szmielew, Wanda.
From affine to Euclidean geometry.

Translation of: Od geometrii afinicznej do euklidesowej.
Bibliography: p.
Includes index.
1. Geometry. 2. Geometry, Affine. I. Title.
QA445. S9513 1982 516 82-20415
ISBN 90-277-1243-3 (Reidel)

Cover design: Zygmunt Ziemka

Translated from the original Polish
Od geometrii afinicznej do euklidesowej. Rozważania nad aksjomatyką
Państwowe Wydawnictwo Naukowe, Warszawa 1981.

Published by PWN-Polish Scientific Publishers, 00-251 Warszawa, Miodowa 10, Poland
in co-publication with
D. Reidel Publishing Company, P. O. Box 17, 3300 AA Dordrecht, Holland.

Distributors for Albania, Bulgaria, Chinese People's Republic, Cuba, Czechoslovakia, German Democratic Republic, Hungary, Korean People's Democratic Republic, Mongolia, Poland, Rumania, the U.S.S.R., Vietnam, and Yugoslavia
Ars Polona,
Krakowskie Przedmieście 7, 00-068 Warszawa, Poland.

Distributors for the U.S.A. and Canada
Kluwer Boston, Inc.,
190 Old Derby Street, Hingham, MA 02043, U.S.A.

Distributors for all remaining countries
Kluwer Academic Publishers Group,
P.O. Box 322, 3300 AH Dordrecht, Holland.

D. Reidel Publishing Company is a member of the Kluwer Group.

Printed in Poland.

TABLE OF CONTENTS

Introduction IX

PART I 1

0. Algebraic structures 1

0.1.—0.3. From loops to commutative fields 1
0.4. Ternary fields. 6
0.5. Strong quasi-fields as ternary fields. 8
0.6. Selected algebraic notions 9

1. Collinearity structures 12

1.1. Collinearity 12
1.2. Lines 14

2. Affine planes 16

2.1. Affine planes as collinearity structures 16
2.2. Parallelity planes 19
2.3. Parallel lines 23
2.4. Parallelograms and trapezoids 26
2.5. Translations and homotheties 29

3. Desarguean and Pappian affine planes 38

3.1. Transitive groups of translations and homotheties . . 38
3.2. Desargues laws 40
3.3. Pappus's law 43
3.4. Translation planes 52
3.5. Homothety planes 58

4. Coordinatization and algebraic representation of affine planes 67

4.0. Algebraic parallelity planes 67
4.1. Coordinatization and its parameters 70
4.2. Construction of the Cartesian coordinate system . . . 72

4.3. Construction of the double loop 74
4.4. Construction of the ternary field and algebraic description of the parallelity relation 78
4.5. Representation theorems 81
4.6. Appendix to § 2.2. 85

PART II 88

5. Midpoint algebras 88

5.1. Groups with the operator $\frac{1}{2}$ 88
5.2. Midpoint algebras 89
5.3. Groups with the operator $\frac{1}{2}$ as midpoint algebras with zero 91
5.4. Vector groups 93

6. Translation planes of characteristic $\neq 2$ 96
6.1. Characteristic $\neq 2$ 96
6.2. Geometric midpoints 100
6.3. Midpoint planes 102
6.4. Representation theorem. 108

7. Algebraic structures with a linear ordering 110
7.1. Linear orders and betweenness relations 110
7.2. (Appendix to § 7.1) The betweenness relation on a set of cardinality $\neq 4$ 120
7.3. Ordered groups 125
7.4. (Appendix to § 7.3) Orderability of a commutative group without cyclic elements 129
7.5. Ordered quasi-fields 132
7.6. Ordered midpoint algebras 136
7.7. Orderings compatible with a midpoint operation . . . 142

8. Ordered affine planes 146
8.1. The betweenness relation on an affine plane and the Pasch axiom 146
8.2. Ordered translation planes 153
8.3. Representation theorem. 156

PART III 162

9. Euclidean planes 162

9.1. Desarguean Euclidean planes 162
9.2. Equi-orthogonal basis 169
9.3. Coordinatization 175
9.4. Pappian Euclidean planes 180
9.5. Equidistance and perpendicularity. 183

Bibliography 186

Bibliography of the works of Wanda Śzmielew 188

List of algebraic symbols 190

List of geometric symbols 191

Index 193

INTRODUCTION

Non omnis moriar

This small monograph was written in rather unusual circumstances. Though it is a result of many years of deliberation, the idea of writing the book occurred to the author in the autumn of 1974, when she was no longer able to carry on her university duties. During the next two years she worked intensively. The monograph soon began to live its own life, to develop beyond the initial framework. But, in May 1976 it became clear that Wanda Szmielew had neither enough time nor strength to complete her work. At the end of May she gave me her typescript. It contained the second version of Part I and the first version of Parts II and III. During the next, and actually the last, three months of her life, she succeeded in giving me detailed remarks concerning Part I and to help with some improvements of the remaining two parts. Afterwards, I tried to change only what was absolutely necessary, so as to preserve as far as possible the original text. According to the author's wish, if I had to change or add anything, I pointed it out by writing "M.M." and referring to the sources if such existed. I used the author's notes as well as the material from her monographic lecture and from the seminar she had been conducting for many years. As a rule, I added only what she herself had intended to add; some fragments of § 7 and the end of § 8 are an exception which—in my opinion—was necessary. I included in Part I a result obtained by L. Włodarski at the end of 1976; therefore some small editorial changes in § 3.3 had to be made (theorems $3.3.2^1$–3.3.4 were added, and consequently the conjunction Papp $\wedge$ papp was replaced by Papp).

Let us now turn to the conception of the book.

Alfred Tarski was the first to find a suitable axiom system of Euclidean geometry, making it possible to develop its foundations in the contemporary meaning of the word (see Tarski [59]). His axiom system describes the class of Euclidean planes over arbitrary real closed ordered fields, Euclidean planes being treated as relational structures with two relations, betweenness and equidistance.

Wanda Szmielew had been Tarski's student. She always considered herself as his continuator in the general treatment of elementary theories[1] and in her love of beautiful axiom systems. But, in course of time, she became more geometer than logician; she began to think about an axiom system which would make it possible to examine very weak structures and to pass step by step to stronger, more special ones, by adding new primitive notions or new axioms. She began to look for an axiom system which would be "natural and acceptable to people" (her expression), and would show clearly the relationships between geometry and algebra. These two aspects, geometric and algebraic, have their counterparts in the structure of the book: both parts I and II start with algebra (the titles show which chapters are of algebraic character), then classes of geometric structures are studied, and finally these two kinds of structures, algebraic and geometric, are connected by means of representation theorems.

Part I contains a full exposition of affine geometry, including very weak theories, starting with the theory of affine planes over ternary fields. Readers who are not interested in such a weak geometry, may omit ternary fields and start reading from quasi-fields. The approach is new. Previously affine planes were treated either as incidence structures, consisting of two universes (of points and lines) and incidence relation, or as collinearity structures, consisting of one universe and collinearity relation. Here, affine geometry is based on the new primitive notion of parallelity, which is a quaternary relation for points. One of the advantages of Wanda Szmielew's axiom system is its $\forall\exists$-character,[2] which implies the summability of the class of models. If parallelity is eliminated by expressing the axioms in terms of collinearity, then the resulting theory, though definitionally equivalent to the original one, has no $\forall\exists$-axiom system. This interesting result (which simultaneously shows that summability of the class of models is not invariant under definitional

[1] An elementary theory is a theory formalized within elementary logic, i.e. first-order predicate calculus (comp. Tarski [59]).

[2] A prenex sentence is of $\forall\exists$-character or simply a $\forall\exists$-sentence (or a universal-existential sentence) iff its universal quantifiers (if any) precede all its existential quantifiers (if any).

equivalences) is presented in the Appendix to § 2.2. This Appendix, as well as the other two, may be omitted by those readers who are not particularly interested in special topics.

Desargue's axioms express in elementary language the transitivity of the translation group and homothety groups. Hence, translation planes and homothety planes are described in three different ways: first, in terms of transformations (i.e. non-elementarily), then (elementarily) by means of Desargues' axioms, and finally, by means of the representation theorem, in terms of algebraic structures.

The style of Part II differs from that of Part I. To some extent, this was the author's intention; on the other hand, this part was in a less finished state than the other two. It concerns the translation planes of characteristic $\neq 2$, and contains many results of the author's own, including the characterization of this class of planes by means of midpoint algebra (§ 6.3), a very simple axiom system of linear order (§ 7.1), results concerning ordered midpoint algebra (§§ 7.6 and 7.7) and their application to the theory of ordered homothety planes (§ 8). The introduction of the midpoint algebra results in both minor Desargues' axiom and characteristic $\neq 2$. Appendix to § 7.1 contains a result obtained by Z. Piesyk. The author intended to give two appendices to § 5, containing results on midpoint algebras obtained by E. Doraczyńska (see Doraczyńska [74]) and by K. Hauschild and W. Rautenberg (see Hauschild & Rautenberg [74]). She also planned to replace fields by algebraic structures with the midpoint operation $\oplus$ instead of addition $+$. The book concerns, in principle, elementary theories; some fragments of Part II make an exception, since they concern the continuity axiom and Archimedes axiom (§§ 7 and 8).

Part III, concerning Euclidean geometry, is the last and the shortest part of the monograph. The language of affine geometry is here extended to the language of Euclidean geometry by adding a ternary predicate which denotes the relation of equi-ortogonal basis, $\boldsymbol{O}$ (instead of the traditionally used predicate of equidistance); the axiom system of affine geometry is extended to an axiom system of Euclidean geometry by choosing simple properties of the basis and its connections with parallelity and midpoint operation. This new approach leads to the class of Euclidean planes over arbitrary (for the first time not necessarily commutative!) 2-formally real fields. The starting point is here the class of Desarguean planes, thus the midpoint operation is definable in terms of parallelity. Nevertheless, this operation is not eliminated, since, in addition to other advantages, such an approach shows analogies between geometric and algebraic structures. I know that the author intended to go on with Part III, to add the axiom of segment construction, and thus to restrict the class of fields to that of Pythagorean fields. Then, probably, she would have extended the structure $(\boldsymbol{S}, \oplus, \|, \boldsymbol{O})$ by the betweenness relation $\boldsymbol{B}$, so as to obtain ordered Euclidean plane. The class of such planes is probably definitionally equivalent to that studied by the author in [74].

Though Wanda Szmielew constantly makes use of the notion of definitional equivalence of structures,[3] nevertheless she deliberately does not embark upon metamathematical discussion concerning this notion. She speaks about definitional equivalence very suggestively, trying to make it as natural for the reader as the rules of inference, which are always used but never formulated in mathematical books. In her opinion (which I entirely share) many mathematical theories would become much clearer if some connections between axiom systems or classes of structures were expressed by means of this notion. It seems natural to identify definitionally equivalent classes of structures, if we treat, as the author does, mathematical objects as reality, which may be described in one language or another one. According to her wish, I draw attention to the notion of definitional equivalence only here, in the introduction. I restrict myself to definitional equivalence of the first order. When admitting non-elementary definitions, one clearly obtains a more general notion.

Let us consider two elementary theories T and T′ with predicates $\boldsymbol{P}$ and $\boldsymbol{P}'$, respectively (for simplicity we assume that each of T and T′ is a theory with only one predicate). These two theories are definitionally equivalent if and only if there exist two elementary definitions, a definition σ of $\boldsymbol{P}$ in terms of $\boldsymbol{P}'$ and a definition σ' of $\boldsymbol{P}'$ in terms of $\boldsymbol{P}$, such that the sets of consequences $\mathrm{Cn}(\mathsf{T} \cup \{\sigma'\})$ and $\mathrm{Cn}(\mathsf{T}' \cup \{\sigma\})$ coincide. Instead of theories, one may consider the classes of their models, i.e. speak about definitional equivalence of two classes of relational structures,

---

[3] This notion (a particular case of mutual interpretability) was introduced by Tarski; see L. Henkin, J. D. Monk, A. Tarski, Cylindric Algebras, 1971.

$\mathscr{C}$ and $\mathscr{C}'$.[4] For simplicity let us assume that the elements of each class are structures with only one relation, $\boldsymbol{P}$ or $\boldsymbol{P}'$ respectively. The classes $\mathscr{C}$ and $\mathscr{C}'$ are definitionally equivalent if, and only if, there is a bijection $\Phi:\mathscr{C} \to \mathscr{C}'$ which assigns to every structure $\mathfrak{A} = (\boldsymbol{S}, \boldsymbol{P}) \in \mathscr{C}$ a structure $\mathfrak{A}' = (\boldsymbol{S}, \boldsymbol{P}') \in \mathscr{C}'$ (with the same universe $\boldsymbol{S}$) such that $\boldsymbol{P}$ is elementarily definable by means of $\boldsymbol{P}'$ and conversely, and moreover the definitions can be chosen in such a way that they are formally the same for all the pairs of corresponding structures involved. Thus, there exist two mutually inverse definitional functors $\Phi:\mathscr{C} \to \mathscr{C}'$ and $\Phi':\mathscr{C}' \to \mathscr{C}$, i.e. functors generated by some elementary definitions of $\boldsymbol{P}$ in terms of $\boldsymbol{P}'$ and conversely. As is well known, two elementary theories T and T' are definitionally equivalent if, and only if, their classes of models are definitionally equivalent. Definitional equivalence of two classes of structures frequently appears in this book (see e.g. § 5.3 or § 7.1). The author also speaks about pairs of definitionally equivalent structures, as structures which have a common definitional expansion (see § 2.2). Hence, on the one hand, corresponding structures of two definitionally equivalent classes are definitionally equivalent, and, on the other hand, if two structures $\mathfrak{A}$ and $\mathfrak{A}'$ are definitionally equivalent then the classes $\{\mathfrak{A}\}$ and $\{\mathfrak{A}'\}$ are definitionally equivalent.

The notation should not present any difficulty. However, let me point out that the symbols $\vee$, $\wedge$, $\sim$, $\Rightarrow$ stand for disjunction, conjunction, negation and implication, respectively, while the symbol $\mapsto$ is used for inference. The symbols $\forall$ and $\exists$ stand for the universal and existential quantifiers.

If there are any omissions in the bibliography, I beg the authors' forgiveness. It would probably have been much more comprehensive if Wanda Szmielew herself had been able to compile it.

Finally, I should like to express warm gratitude—on behalf of the author and myself—to dr Stefan Madajewicz, Wanda Szmielew's medical attendant, whose unremitting medical care enabled her to write this book. I should also like to offer our sincere thanks to dr Marek Kordos for his advice and help in editing the book and preparing it for the press.

Warsaw, May 3, 1977. *Maria Moszyńska*

[4] See footnote 3.

PART I

CHAPTER 0

ALGEBRAIC STRUCTURES

0.1–0.3. FROM LOOPS TO COMMUTATIVE FIELDS

The simplest algebraic structure[1] is a *groupoid* $(G, +)$ which consists of an arbitrary set G and arbitrary binary operation $+$ on G.[2] A groupoid with a distinguished element, $(G, 0, +)$, is said to be a *loop* if for every $a \in G$

$$a+0 = 0+a = a,$$

and for any $a, b \in G$ each of two equations

$$a+x = b \quad \text{and} \quad x+a = b$$

has a unique solution in G. The operation $+$ is referred to as *addition*, and the element 0 as the *zero* of the loop $(G, 0, +)$.

The above conditions describing a loop can be replaced by the following six axioms:[3]

Ax 0.1.1. $a+0 = a.$

Ax 0.1.1′. $0+a = a.$

Ax 0.1.2. $\forall ab \exists x \;\; a+x = b.$

Ax 0.1.2′. $\forall ab \exists x \;\; x+a = b.$

Ax 0.1.3. $a+x = a+x' \Rightarrow x = x'.$

Ax 0.1.3′. $x+a = x'+a \Rightarrow x = x'.$

---

[1] We omit the structures with unary operations.

[2] We use here the additive notation, but—as was mentioned above—the symbol "+" denotes an arbitrary binary operation.

[3] In general, in all the axioms, theorems, and definitions we shall omit the universal quantifiers which should be placed in front of a formula to bind all the free variables occuring in it.

If the operation $+$ is associative, then the loop $(G, 0, +)$ is said to be a *group*. Thus, groups are just those loops which satisfy the axiom

Ax 0.1.4. $(a+b)+c = a+(b+c)$.

The axiom system for a group can be simplified as follows. Let us consider a particular case of the axiom 0.1.2:

Th 0.1.2weak. $\forall a\, \exists x \;\; a+x = 0$.

We shall show that all the axioms of a group are derivable from the axioms

0.1.1, 0.1.2weak, 0.1.4.

First, let us derive the following three auxiliary statements:

Th 0.1.5. $a+b = 0 \Rightarrow b+a = 0$.

Proof. Let

(1) $a+b = 0$.

By 0.1.2weak, for some $x \in G$

(2) $b+x = 0$.

From (1) and (2), using 0.1.1 and 0.1.4 we obtain in turn

$$b+(a+b) = b,$$

$$b+a = (b+a)+(b+x) = (b+(a+b))+x = b+x = 0. \blacksquare$$

Th 0.1.6. $0+a = a+0$.

Proof. By 0.1.2weak and 0.1.5, for some $x \in G$

$$a+x = 0 \quad \text{and} \quad x+a = 0,$$

and thus, by 0.1.4,

$$0+a = (a+x)+a = a+(x+a) = a+0. \blacksquare$$

Th 0.1.2′weak. $\forall a\, \exists x \;\; x+a = 0$.

Proof. by 0.1.2weak and 0.1.5. ■

We have now the following obvious inferences (starting from the second one—by 0.1.4):

$$\{0.1.6, \quad 0.1.1\} \mapsto 0.1.1',$$

$$\{0.1.2\text{weak}, \quad 0.1.1'\} \mapsto 0.1.2,$$

$\{0.1.2\text{'weak},\ 0.1.1\} \mapsto 0.1.2'$,

$\{0.1.2'\text{weak},\ 0.1.1'\} \mapsto 0.1.3$,

$\{0.1.2\text{weak},\ 0.1.1\} \mapsto 0.1.3'$.

In conclusion, a group $(G, 0, +)$ is described by the axiom system $\{0.1.1,\ 0.1.2\text{weak},\ 0.1.4\}$. Adding the axiom

Ax 0.1.7. $a+b = b+a$,

which postulates the commutativity of addition, we distinguish *commutative groups*. Hence, the set of sentences

$$\text{AX } 0.1 = \{0.1.1,\ 0.1.2\text{weak},\ 0.1.4,\ 0.1.7\}$$

is an axiom system for commutative groups.

In an arbitrary group, *subtraction* is introduced in the familiar way. Thus, the element $-a$ *opposite to* a is defined as the unique solution of the equation $a+x = 0$, and so of the equation $x+a = 0$ as well (see 0.1.5). We have therefore

Def 0.1.8.
Th $a+(-a) = (-a)+a = 0$.

Def 0.1.9. $a-b = a+(-b)$.

In the multiplicative notation, any loop and in particular a group is a groupoid $(G, 1, \cdot)$. The operation $\cdot$ is referred to as the *multiplication* and the element 1 as the *unit* of a loop. We modify the axiom system by replacing all the signs "+" and "0" by the signs "$\cdot$" and "1" respectively. Of course, in a group, instead of $-a$ we have the element a^{-1} *inverse to a.*

A groupoid with two distinguished elements, $(G, 0, 1, \cdot)$, is said to be a *multiplicative loop with zero* if

(i) $0 \neq 1$,

(ii) $(G - \{0\}, 1, \cdot)$ is a loop,

(iii) $a \cdot 0 = 0$ and $0 \cdot a = 0$ for every $a \in G$.

The elements 0 and 1 are referred to as, respectively, the *zero* and *unit* of the loop $(G, 0, 1, \cdot)$.

The conditions (i)–(iii) can be replaced by the following axioms:

Ax 0.2.0. $0 \neq 1.$

Ax 0.2.1. $a \cdot 1 = a.$

Ax 0.2.1′. $1 \cdot a = a.$

Ax 0.2.2. $a \neq 0 \Rightarrow \forall b \, \exists x \;\; a \cdot x = b.$

Ax 0.2.2′. $a \neq 0 \Rightarrow \forall b \, \exists x \;\; x \cdot a = b.$

Ax 0.2.3. $a \neq 0 \wedge a \cdot x = a \cdot x' \Rightarrow x = x'.$

Ax 0.2.3′. $a \neq 0 \wedge x \cdot a = x' \cdot a \Rightarrow x = x'.$

Ax 0.2.4. $a \cdot 0 = 0.$

Ax 0.2.4′. $0 \cdot a = 0.$

In particular, if the loop $(G - \{0\}, 1, \cdot)$ is a group, i.e. if the operation $\cdot$ is associative, then the loop with zero, $(G, 0, 1, \cdot)$, is said to be a *multiplicative group with zero*. Thus, multiplicative groups with zero are just those multiplicative loops with zero which satisfy the axiom

Ax 0.2.5. $(a \cdot b) \cdot c = a \cdot (b \cdot c).$

Let us consider a particular case of 0.2.2:

Ax 0.2.2weak. $a \neq 0 \Rightarrow \exists x \;\; a \cdot x = 1.$

A discussion similar to that concerning a group $(G, 0, +)$ shows that the set of sentences

AX 0.2 = {0.2.0, 0.2.1, 0.2.2weak, 0.2.4, 0.2.4′, 0.2.5} is an

axiom system for multiplicative groups with zero. Adding to this system the axiom

Ax 0.2.6. $a \cdot b = b \cdot a,$

we distinguish the *commutative multiplicative groups with zero*.

We pass to algebraic structures with two operations.[4] A structure $\mathfrak{F} = (F, 0, 1, +, \cdot)$ is said to be a *double loop*, if its substructures $\mathfrak{F}_+ = (F, 0, +)$ and $\mathfrak{F}_\cdot = (F, 0, 1, \cdot)$ are respectively an additive[5] loop and a multiplicative loop with zero. If, moreover, the loop $\mathfrak{F}_+$ is a commutative group, and simultaneously the multiplication is left distribu-

[4] At this moment, we change the symbol "G" into "F" for a universe.

[5] I.e. the loop $(F, 0, +)$, as a substructure of the double loop $(F, 0, 1, +, \cdot)$, is now referred to as an *additive loop*.

tive with respect to addition, then the double loop $\mathfrak{F}$ is referred to as a *left quasi-field.*[6] This left distributivity is expressed by

Ax 0.3.1. $a\cdot(x+y) = a\cdot x + a\cdot y,$

and any left quasi-field is described by the axiom system

$$\text{AX } 0.1 \cup \{0.2.0, 0.2.1, \ldots, 0.2.4, 0.2.1', \ldots, 0.2.4'\} \cup \{0.3.1\}.$$

As consequences of these axioms we obtain the following theorems:

Th 0.3.2. $a\cdot(-b) = -(a\cdot b).$

P r o o f. By 0.3.1, 0.1.8, and 0.2.4,

$$a\cdot(-b)+a\cdot b = a\cdot\big((-b)+b\big) = a\cdot 0 = 0. \;\blacksquare$$

Th 0.3.3. $(-1)\cdot(-1) = 1.$

P r o o f. By 0.3.2, 0.2.1, and the (additive) group axioms,

$$(-1)\cdot(-1) = -\big((-1)\cdot 1\big) = -(-1) = 1. \;\blacksquare$$

Th 0.3.4. $a\cdot(x-y) = a\cdot x - a\cdot y.$

P r o o f. By 0.1.9, 0.3.1, and 0.3.2,

$$a\cdot(x-y) = a\cdot x + a\cdot(-y) = a\cdot x - a\cdot y. \;\blacksquare$$

Thus, in a left quasi-field multiplication is left distributive with respect to subtraction as well.

If, instead of the left distributivity, we assume the right distributivity of multiplication with respect to addition, i.e. if we replace the axiom 0.3.1 by

Ax 0.3.1′. $(x+y)\cdot a = x\cdot a + y\cdot a,$

then we obtain a description of a right quasi-field.

Any structure $\mathfrak{F} = (F, 0, 1, +, \cdot)$ which is a left and right quasi--field simultaneously, is said to be a *two-sided quasi-field.* A two-sided quasi-field with an associative multiplication (vide 0.2.5) is a *field.*[7]

[6] In the literature, there are some additional conditions to be satisfied by a left quasi-field (compare the footnote 10). Often the terms "left" and "right" are interchanged.

[7] In the literature there is often the additional condition 0.2.6 to be satisfied by a field; non-commutative fields are called skew-fields (M.M.).

Thus the substructure $\mathfrak{F}.$ of a field $\mathfrak{F}$ is a multiplicative group with zero, and the axiom system for fields reduces to the set

$$\text{AX } 0.1 \cup \text{AX } 0.2 \cup \{0.3.1, 0.3.1'\}.$$

Adding the axiom 0.2.6, we distinguish the *commutative fields* from among all the fields. Of course, in the presence of 0.2.6 the axioms 0.2.4′ and 0.3.1′ may be omitted.

0.4. TERNARY FIELDS[8]

We now pass to very special algebraic structures, which are closely related to affine planes. Consider an arbitrary structure $(F, 0, 1, \mathbf{T})$ which consists of a set F, two distinguished elements $0,1 \in F$, and a ternary operation $\mathbf{T}$ on F. Such a structure is said to be a *ternary field*, if the following conditions are satisfied:

Ax 0.4.0. $0 \neq 1.$

Ax 0.4.1. $\mathbf{T}(a10) = a.$

Ax 0.4.2. $\mathbf{T}(1a0) = a.$

Ax 0.4.3. $\mathbf{T}(a0b) = b.$

Ax 0.4.4. $\mathbf{T}(0ab) = b.$

Ax 0.4.5. $\forall uab\, \exists v\; \mathbf{T}(uav) = b.$

Ax 0.4.6. $\mathbf{T}(uav) = \mathbf{T}(uav') \Rightarrow v = v'.$

Ax 0.4.7. $a \neq a' \Rightarrow \forall bb' \exists uv\; (\mathbf{T}(uav) = b \wedge \mathbf{T}(ua'v) = b').$

Ax 0.4.8. $u \neq u' \Rightarrow \forall vv'\, \exists a\; \mathbf{T}(uav) = \mathbf{T}(u'av').$

Ax 0.4.9. $\mathbf{T}(uav) = \mathbf{T}(u'av') \wedge \mathbf{T}(ua'v) = \mathbf{T}(u'a'v')$
$\Rightarrow a = a' \vee u = u'.$

By 0.4.6, axiom 0.4.9 can be strengthened to

Th 0.4.10. $a \neq a' \wedge \mathbf{T}(uav) = \mathbf{T}(u'av') \wedge \mathbf{T}(ua'v) = \mathbf{T}(u'a'v')$
$\Rightarrow u = u' \wedge v = v'.$

To discover the intuition behind these axioms, regard the formula

(i) $\mathbf{T}(uxv) = y$

as the equation of a straight line $(y = u \cdot x + v)$. Then the axioms 0.4.5–0.4.9 become algebraic analogues of the following geometric

[8] In the Polish edition, ternary fields are referred to as affinoids (M.M.).

statements: two distinct points determine a unique line (0.4.7 and 0.4.10); there is a unique line which passes through a given point and has a given direction (0.4.5 and 0.4.6); two non-parallel lines have a point in common (0.4.8).

Let us put aside these geometric intuitions and look at ternary field as an algebraic structure.

Since **T** is a ternary operation, it may have three inverse operations. The axioms 0.4.5 and 0.4.6 say that **T** has an inverse operation with respect to the third argument. Let us write it down again using other variables:

Th 0.4.11. $\forall abc\,\exists x\ \mathbf{T}(abx) = c.$

Th 0.4.12. $\mathbf{T}(abx) = \mathbf{T}(abx') \Rightarrow x = x'.$

We are going to show that **T** has (almost everywhere defined) inverse operations with respect to the remaining two arguments as well:

Th 0.4.13. $a \neq 0 \Rightarrow \forall bc\,\exists x\ \mathbf{T}(axb) = c.$

Proof. By 0.4.8 and 0.4.4,

$$a \neq 0 \Rightarrow \forall bc\,\exists x\ \mathbf{T}(axb) = \mathbf{T}(0xc) = c. \blacksquare$$

Th 0.4.14. $a \neq 0 \wedge \mathbf{T}(axb) = \mathbf{T}(ax'b) \Rightarrow x = x'.$

Proof. Let

$$\mathbf{T}(axb) = \mathbf{T}(ax'b) = c.$$

By 0.4.4,

$$\mathbf{T}(axb) = \mathbf{T}(0xc) \quad \text{and} \quad \mathbf{T}(ax'b) = \mathbf{T}(0x'c),$$

which, together with $a \neq 0$, by 0.4.9 implies $x = x'$. ■

Th 0.4.15. $a \neq 0 \Rightarrow \forall bc\,\exists x\ \mathbf{T}(xab) = c.$

Proof. By 0.4.7 and 0.4.3,

$$a \neq 0 \Rightarrow \forall bc\,\exists xz\ \big(\mathbf{T}(xaz) = c \wedge z = \mathbf{T}(x0z) = b\big). \blacksquare$$

Th 0.4.16. $a \neq 0 \wedge \mathbf{T}(xab) = \mathbf{T}(x'ab) \Rightarrow x = x'.$

Proof. By 0.4.3,

$$\mathbf{T}(x0b) = \mathbf{T}(x'0b),$$

which, together with the assumption, by 0.4.9 implies $x = x'$. ■

By means of the operation **T**, addition and multiplication can be introduced as follows:

Def 0.4.17. $a+b = \mathbf{T}(1ab)$.

Def 0.4.18. $a \cdot b = \mathbf{T}(ab0)$.

Theorem 0.4.19. *The structure* $(F, 0, 1, +, \cdot)$ *is a double loop.*

Proof. Applying 0.4.0–0.4.4 and 0.4.11–0.4.16, we can easily verify all the axioms of an additive loop $(F, 0, +)$ and all the axioms of a multiplicative loop with zero, $(F, 0, 1, \cdot)$. ∎

As we have seen, by regarding formula (i) as the equation of a line, one gets a geometric interpretation of axioms 0.4.5–0.4.9. Thus, it is natural to ask, whether the formula

$$\mathbf{T}(uxv) = u \cdot x + v,$$

and hence also the equivalent formula

$$\text{(ii)} \qquad \mathbf{T}(uxv) = \mathbf{T}(1\mathbf{T}(ux0)\, v),$$

are consequences of 0.4.0–0.4.9. The answer is negative: there exist ternary fields which do not satisfy (ii).[9] In conclusion, the operation **T** cannot be (in this way) defined in terms of sum and product.

0.5. STRONG QUASI-FIELDS AS TERNARY FIELDS

We are going to establish a relationship between ternary fields and left quasi-fields. A left quasi-field $\mathfrak{F} = (F, 0, 1, +, \cdot)$ is said to be *strong* if it satisfies the following additional axiom

Ax 0.2.2str. $a_1 \neq a_2 \Rightarrow \forall b_1 b_2 \exists x \;\; a_1 \cdot x + b_1 = a_2 \cdot x + b_2$.

This axiom is, of course, stronger than 0.2.2 ($a_1 = a, a_2 = b_1 = 0$, $b_2 = b$).[10] Notice that the analogous stronger form of 0.2.2′ is a consequence of the axioms for left quasi-fields, since it can be derived from 0.2.2′ and the axioms for additive groups, by means of the left distributivity law.

---

[9] See Pickert [75], p. 293.

[10] In the literature, such a strong left quasi-field is referred to as left quasi-field (or left distributive Cartesian group). Pickert in [75] gives its axiom system; (see Pickert [75], p. 90 and last three lines on p. 92).

Given an arbitrary double-loop $\mathfrak{F} = (F, 0, 1, +, \cdot)$, let us define a ternary operation **T**:

Def 0.5.1. $\mathbf{T}(abc) = a \cdot b + c.$

We have

Th 0.5.2. $b+c = \mathbf{T}(1bc)$ and $a \cdot b = \mathbf{T}(ab0).$

Consider the structure $\mathfrak{T}(\mathfrak{F}) = (F, 0, 1, \mathbf{T})$. By 0.5.1 and 0.5.2, the structure $(F, 0, 1, +, \cdot, \mathbf{T})$ is a definitional expansion of both $\mathfrak{F}$ and $\mathfrak{T}(\mathfrak{F})$. It is easily seen that if the loop $\mathfrak{F}$ is a strong left quasi-field, then $\mathfrak{T}(\mathfrak{F})$ is a ternary field (of the quasi-field $\mathfrak{F}$), and the axiom 0.2.2str of a quasi-field is needed just to verify the axiom 0.4.8 for ternary fields.

Thus, in fact, the class of ternary fields contains all the strong left quasi-fields, in particular the (commutative) fields.

Let the symbols

$$\mathbf{TF}, \mathbf{QF}, \mathbf{F}, \mathrm{com}\,\mathbf{F}$$

denote respectively the class of ternary fields, strong left quasi-fields, fields, and commutative fields.[11] We have

$$\mathbf{TF} \underset{\mathrm{Def}}{\supseteq} \mathbf{QF} \supseteq \mathbf{F} \supseteq \mathrm{com}\,\mathbf{F}.$$

0.6. SELECTED ALGEBRAIC NOTIONS

Of course, we shall use the well known algebraic notions. Let us recall only some of them.

Let $\mathfrak{F} = (F, 0, 1, +, \cdot)$ be a double loop. We define bijections

$$\sigma^v : F \to F \quad \text{and} \quad {}^u\psi : F \to F$$

as follows:

Def 0.6.1. (i) $\sigma^v(x) = x+v$ for $v \in F$

(ii) ${}^u\psi(x) = u \cdot x$ for $u \in F - \{0\}$.

Let $\mathfrak{G} = (G, 0, +)$ be a group, We define the bijection

$$\mu : G \to G$$

[11] In the Polish edition the symbol **A** is used instead of **TF** (comp. footnote 8) (M.M.).

as follows:

Def 0.6.2. $\mu(x) = -x.$

The *Cartesian square* of the group $\mathfrak{G}$ is a structure

$$\mathfrak{G}^2 = (G^2, 0, +),$$

with

$$0 = (0, 0) \quad \text{and} \quad (a_1, a_2)+(b_1, b_2) = (a_1+b_1, a_2+b_2).$$

Evidently $\mathfrak{G}^2$ is a group.

Let $\mathfrak{F} = (F, 0, 1, +, \cdot) \in \mathbf{F}$. The right Cartesian (2-dimensional) linear space over $\mathfrak{F}$ is defined to be a structure

$$\underline{\mathfrak{B}}^2(\mathfrak{F}) = (\mathfrak{F}^2_+, \cdot),$$

where $\cdot$ is the right multiplication of a point by a scalar:

$$(a_1, a_2)\cdot u = (a_1 \cdot u, a_2 \cdot u).$$

The left Cartesian linear space can be defined analogically.

Now, let $\mathfrak{F} \in \mathbf{F}$ and let $a_1, b_1, a_2, b_2 \in F$. Consider the following three conditions:

(i) $a_1 = 0 \wedge a_2 = 0 \vee \exists u\, (u \cdot a_1 = b_1 \wedge u \cdot a_2 = b_2)$,

(ii) $\exists uv\, \big((u, v) \neq (0, 0) \wedge u \cdot a_1 = v \cdot b_1 \wedge u \cdot a_2 = v \cdot b_2\big)$,

(iii) $\exists uv\, \big((u, v) \neq (0, 0) \wedge a_1 \cdot u = a_2 \cdot v \wedge b_1 \cdot u = b_2 \cdot v\big)$.

Th 0.6.3. (i) $\rightleftarrows$ (ii).

Proof. Obviously (i) $\Rightarrow$ (ii). Assume (ii). If $v = 0$, then $u \neq 0$ and $u \cdot a_1 = u \cdot a_2 = 0$, thus $a_1 = a_2 = 0$. If $v \neq 0$, then for $u' = v^{-1} \cdot u$ we have $u' \cdot a_i = (v^{-1} \cdot u) \cdot a_i = v^{-1} \cdot (u \cdot a_i) = v^{-1} \cdot (v \cdot b_i) = (v^{-1} v) \cdot b_i = b_i$ for $i = 1, 2$. Hence (ii) $\Rightarrow$ (i). ■

Th 0.6.4. (i) $\rightleftarrows$ (iii).

Proof of (i) $\Rightarrow$ (iii). The conjunction

(1) $a_1 \cdot u = a_2 \cdot v \wedge b_1 \cdot u = b_2 \cdot v$

is satisfied

by $u = 1, v = 0$ if $a_1 = 0$ and $b_1 = 0$,

by $u = 0, v = 1$ if $a_2 = 0$ and $b_2 = 0$,

by $u = b_1^{-1}, v = b_2^{-1}$ if $a_1 = 0, a_2 = 0$ and $b_1, b_2 \neq 0$.

Let $a_1 \neq 0$. By (i), there is a u satisfying

(2) $\qquad u \cdot a_1 = b_1, \quad u \cdot a_2 = b_2.$

Let $u' = a_1^{-1} \cdot a_2$ and $v' = 1$. Then

$$a_1 \cdot u' = a_1 \cdot (a_1^{-1} \cdot a_2) = (a_1 \cdot a_1^{-1}) \cdot a_2 = a_2 \cdot v'$$

and

$$\begin{aligned} b_1 \cdot u' &= (u \cdot a_1) \cdot (a_1^{-1} \cdot a_2) = u \cdot (a_1 \cdot a_1^{-1}) \cdot a_2 = u \cdot a_2 \\ &= b_2 \cdot v'. \end{aligned}$$

Thus u', v' satisfy (1). When $a_2 \neq 0$, the proof is similar. Hence (i) $\Rightarrow$ (iii).

Now, assume (iii). Let $(u, v) \neq (0, 0)$ satisfy (1). If $a_1 \neq 0$, then $v \neq 0$. Let $u' = b_1 \cdot a_1^{-1}$. Then

$$u' \cdot a_1 = (b_1 \cdot a_1^{-1}) \cdot a_1 = b_1 \cdot (a_1^{-1} \cdot a_1) = b_1$$

and

$$\begin{aligned} (u' \cdot a_2) \cdot v &= (b_1 \cdot a_1^{-1}) \cdot (a_2 \cdot v) = (b_1 \cdot a_1^{-1}) \cdot (a_1 \cdot u) \\ &= b_1 \cdot (a_1^{-1} \cdot a_1) \cdot u = b_1 \cdot u = b_2 \cdot v, \end{aligned}$$

whence, since $v \neq 0$, it follows that

$$u' \cdot a = b_2 .$$

When $a_2 \neq 0$, the proof is similar. Hence (iii) $\Rightarrow$ (i). ∎

By 0.6.3 and 0.6.4, the proportionality of (a_1, a_2) and (b_1, b_2) in a left linear space ((i), (ii)) is equivalent to the proportionality of (a_1, b_1) and (a_2, b_2) in a right linear space ((iii)).

Let us notice that in the proofs of both 0.6.3 and 0.6.4 only associativity is needed (we do not use distributivity).

CHAPTER 1

COLLINEARITY STRUCTURES

Let us start with the notion of collinearity structure, which is more general than the notion of affine plane. The collinearity will be defined as an arbitrary ternary equivalence, and lines as its equivalence classes.[1]

1.1. COLLINEARITY

Given a set S and a relation $L \subset S \times S \times S$, the pair (S, L) is said to be a *collinearity structure* if it satisfies the following two axioms:[2,3]

Ax 1.1.1. $a = b \vee a = c \vee b = c \Rightarrow L(abc)$.

Ax 1.1.2. $a \neq b \wedge L(abp) \wedge L(abq) \wedge L(abr) \Rightarrow L(pqr)$.

Then the elements of S are called points, and the relation L is referred to as the *collinearity*.

The axiom 1.1.1 postulates the *reflexivity*, the axiom 1.1.2 gives a kind of *transitivity*;[4] both together imply the *symmetry*:

Th 1.1.3. $L(abc) \Rightarrow L(bac) \wedge L(acb)$.

(If, in the consequent of this implication, both transpositions of adjacent elements are taken into account, then the remaining three permutations of the triple (abc) can be omitted).

---

[1] The notion of ternary equivalence was introduced by Bachmann in [59] pp. 64, 65. Both of our axiom systems (see § 1.1) are simpler than analogous Bachmann's axiom systems.

[2] To simplify notation, we use the same letter "L" twice: first as the term of a primitive relation of our structure, and then as the corresponding predicate of the elementary language of logic in which the axioms describing this structure are formulated.

[3] See footnote 3 in chapter 0.

[4] For ternary relations the notion of transitivity is not uniquely determined. Bachmann treats our axiom 1.1.2′ as the transitivity law.

The axioms 1.1.1 and 1.1.2 can be equivalently replaced by the following three:

Ax 1.1.1′. $L(aba)$.

Ax 1.1.2′. $a \neq b \wedge L(abc) \wedge L(abd) \Rightarrow L(acd)$.

Ax 1.1.3′. $L(abc) \Rightarrow L(bac)$.

The derivation of the new axioms from the original axioms is obvious. By the new axioms we get

$$L(aba) \mid\Rightarrow L(baa), \qquad a \neq b \wedge L(aba) \wedge L(abb) \mid\Rightarrow L(aab),$$

and

$$a \neq b \wedge L(abp) \wedge L(abq) \wedge L(abr) \mid\Rightarrow L(apq) \wedge L(apr) \wedge L(aqr)$$
$$\mid\Rightarrow L(paq) \wedge L(par) \wedge L(aqr) \mid\Rightarrow L(pqr). \blacksquare$$

The following theorem is a kind of converse to axiom 1.1.2.

Th 1.1.4. $p \neq q \wedge L(abp) \wedge L(abq) \wedge L(pqr) \Rightarrow L(abr)$.

P r o o f. By 1.1.1–1.1.3 and 1.1.2′

$$a = b \mid\Rightarrow L(abr), \; a \neq b \wedge p \neq q \wedge L(abp) \wedge L(abq) \wedge L(pqr)$$
$$\mid\Rightarrow p \neq q \wedge L(pqa) \wedge L(pqb) \wedge L(pqr) \mid\Rightarrow L(abr). \blacksquare$$

A structure $(S, S \times S \times S)$ is said to be *trivial*. The additional axiom

Ax dim. $\exists abc \sim L(abc)$

excludes trivial structures. If a collinearity structure is not trivial then we say that its *dimension is at least* 2. By 1.1.1, such a structure consists of at least three distinct points, and two arbitrary points can be completed to a non-collinear triple:

Th 1.1.5. $p \neq q \Rightarrow \exists r \sim L(pqr)$.

P r o o f. By dim, for some a, b, c

$$\sim L(abc).$$

So, if $p \neq q$, then by 1.1.2,

$$\sim L(pqa) \vee \sim L(pqb) \vee \sim L(pqc). \blacksquare$$

We choose the set of sentences

$$\text{AX } 1.1 = \{1.1.1', 1.1.2', 1.1.3', \text{dim}\}$$

as an axiom system for non-trivial collinearity structures.

Adding the axiom

Ax Dim. $\forall abcd\, \exists p\, \big((L(abp) \wedge L(cdp)) \vee (L(acp) \wedge L(bdp)) \vee (L(adp) \wedge L(bcp))\big)$,

we distinguish the structures of dimension 2.

1.2. LINES

Let (S, L) be a collinearity structure. With every pair of points we associate a subset of the universe S:

Def 1.2.0. $[ab] = \{p : L(abp)\}$.

By 1.1.1, we have

Th 1.2.1. $a, b \in [ab]$.

By 1.1.1 and 1.1.5 we get

Th 1.2.2. (i) $a = b \Rightarrow [ab] = S$.

(ii) $a \neq b \Rightarrow [ab] \neq S$.

If $a \neq b$, then the set $[ab]$ is referred to as the *line determined by a and b*, or simply the *line ab*. Lines will be denoted by capital Latin letters. We use the phrases: *the point p lies on the line P* or *the line P passes through the point p*, to express the fact that $p \in P$. We say that *the lines P and Q intersect each other* if $P \cap Q \neq \emptyset$, and that *they intersect in the point a* if $P \cap Q = \{a\}$.

From 1.2.1 we deduce

Th 1.2.3. $\forall P\, \exists ab\, (a \neq b \wedge a, b \in P)$

and

Th 1.2.4. $a \neq b \Rightarrow \exists P\; a, b \in P$.

Setting $P = [ab]$, we can express 1.2.2 as follows:

Th 1.2.5. $p, q, r \in P \Rightarrow L(pqr)$.

Th 1.2.6. $P \subset Q \Rightarrow P = Q$.

Proof. If $P = [pq]$ and $Q = [ab]$, then by 1.2.1 and 1.2.5

$$P \subset Q \wedge r \in Q \mapsto p, q, r \in Q \mapsto L(pqr) \mapsto r \in P.$$

Thus $P \subset Q \mapsto Q \subset P$, whence $P \subset Q \mapsto P = Q$. ∎

Setting $P = [ab]$, we can express 1.1.4 as

Th 1.2.7. $p \neq q \wedge p, q \in P \Rightarrow [pq] \subset P$.

From 1.2.6 and 1.2.7 we deduce

Th 1.2.8. $p \neq q \wedge p, q \in P \Rightarrow [pq] = P$.

By 1.2.8 we have

Th 1.2.9. $p \neq q \wedge p, q \in P \wedge p, q \in Q \Rightarrow P = Q$.

It can be reformulated as follows:[5]

Th 1.2.10. $P = Q \vee P \cap Q = \emptyset \vee \exists p \; P \cap Q = \{p\}$.

Whence, there are at least two distinct points lying on a given line (1.2.3), and there is exactly one line passing through two distinct points (1.2.4 and 1.2.9).

[5] This new version (1.2.10) of 1.2.9 is a basic theorem on the classes of ternary equivalence. Two distinct classes may have a point in common but at most one.

CHAPTER 2

AFFINE PLANES

Traditionally, an affine plane is understood as a structure consisting of two sets, a universe of points and a universe of lines, the one connected with the other by means of the incidence relation.[1] Such an approach is justified when affine planes are considered in connection with projective planes (which are dual with respect to the two universes). Otherwise, it is more natural to define an affine plane as a collinearity structure. Then lines are treated as sets of points, i.e. figures of the plane. Such an approach is also traditional. It is presented in § 2.1. A different approach is presented in § 2.2: instead of collinearity, we take parallelity relation as a primitive notion.

2.1. AFFINE PLANES AS COLLINEARITY STRUCTURES

A non trivial collinearity structure $(\boldsymbol{S}, \boldsymbol{L})$ is said to be an *affine plane* whenever the following two conditions are satisfied:

(i) $p \notin P \Rightarrow \exists Q\,(p \in Q \wedge P \cap Q = \emptyset)$.

(ii) $p \in Q \wedge p \in Q' \wedge P \cap Q = \emptyset \wedge P \cap Q' = \emptyset \Rightarrow Q = Q'$.

The first one says that if a point p does not belong to a line P, then there is at least one line Q passing through p and not intersecting P. The second one says that there is at most one such line Q. Condition (ii) is known as *Euclid's axiom on parallel.*

Conditions (i) and (ii), which are concerned, in fact, with parallel lines, should be replaced by elementary axioms. To make the form ade-

[1] The axioms of incidence for an affine plane can be found e.g. in Lenz [68] p. 141, Lingenberg–Baur [60], p. 97, Pickert [55], p. 10. In Pickert [55], p. 2, the notion of affine plane is preceded by a more general notion of *regular incidence structure.* Our collinearity structure corresponds to a regular incidence structure in which every two distinct points are incident with a line.

quate to the matter, we shall express these axioms not directly in terms of $\boldsymbol{L}$ itself,[2] but in terms of the parallelity relation $\|$, which is defined by means of $\boldsymbol{L}$. Since we look for elementary axioms, the parallelity should be considered as a relation for points and defined by the elementary formula:

Ax 2.1.0. $ab\|cd \Leftrightarrow \big(a = b \vee \forall p\ (\boldsymbol{L}(abp) \Rightarrow \boldsymbol{L}(cdp))$
Def $\qquad \vee \forall p\ (\boldsymbol{L}(abp) \Rightarrow \sim \boldsymbol{L}(cdp))\big).$

In other words

$$ab\|cd \Leftrightarrow (a = b \vee [ab] \subset [cd] \vee [ab] \subset \boldsymbol{S} - [cd]),$$

whence, by means of 1.2.2 and 1.2.6 we get immediately

Th 2.1.0′. $ab\|cd \Leftrightarrow (a = b \vee c = d \vee [ab] = [cd]$
$\qquad \vee [ab] \cap [cd] = \emptyset).$

Any unordered pair of points will be referred to as a *segment*. In view of 2.1.0′, the formula $ab\|cd$ reads: *the segment ab is parallel to the segment cd*.[3] By 2.1.0′, the conditions (i) and (ii) are equivalent to the following two sentences:

(i′) $\quad \sim \boldsymbol{L}(abp) \Rightarrow \exists q\ (ab\|pq \wedge p \neq q).$

(ii′) $\quad \sim \boldsymbol{L}(abp) \wedge p \neq q, r \wedge ab\|pq \wedge ab\|pr \Rightarrow \boldsymbol{L}(pqr).$

Indeed, it suffices to take $P = [ab], Q = [pq], Q' = [pr]$ and apply theorems of § 1.2. Notice that in (i′) the assumption $\sim \boldsymbol{L}(abp)$ can be omitted, because if $\boldsymbol{L}(abp)$ then, by 2.1.0′, the existence of the corresponding point b is a consequence of AX 1.1. For the same reason, in (ii′) it suffices to assume $a \neq b$. Finally, we replace the conditions (i) and (ii) by three: the definitional axiom 2.1.0[4] and the following two axioms on parallelity:[5]

---

[2] It would lead to syntactically complicated sentences, the first of them of $\forall\exists\forall$-type (vide 2.1.1, 2.1.2, and 2.1.0).

[3] Let us recall that in our approach parallelity is a quaternary relation for points. Neither the notation ($ab\|cd$ instead of $\|(abcd)$) nor the way of reading the formula $ab\|cd$ change this fact.

[4] Since the axioms 2.1.1 and 2.1.2 are expressed in terms of $\|$, the definition 2.1.0 must be adjoined to the axiom system. Thus, we refer to 2.1.0 as definitional axiom.

[5] Omitting superfluous assumptions, we simplify syntactically particular axioms but do not strengthen the whole axiom system. We repeat this very often in the sequel.

Ax 2.1.1. $\forall abp \exists q\, (ab||pq \wedge p \neq q)$.

Ax 2.1.2. $a \neq b \wedge ab||pq \wedge ab||pr \Rightarrow \boldsymbol{L}(pqr)$.

Let us notice that the axioms 2.1.0, 2.1.1, and 2.1.2 do not concern the original structure $(\boldsymbol{S}, \boldsymbol{L})$, but its definitional expansion $(\boldsymbol{S}, \boldsymbol{L}, ||)$. This expansion will also be referred to as an *affine plane*. Let

$$\text{AX } 2.1 = \{2.1.0, 2.1.1, 2.1.2\}.$$

The union AX 1.1$\cup$AX 2.1 is an axiom system for the affine planes.

Definition 2.1.0, describing parallelity by means of collinearity, is rather complicated. Indeed, to check if the points $a, b, c, d \in \boldsymbol{S}$ are in the parallelity relation (i.e. $ab||cd$), we must deal (by means of the quantifiers) with the whole universe $\boldsymbol{S}$. On the contrary, collinearity may be defined in terms of parallelity in a very simple way:

Th 2.1.3. $\boldsymbol{L}(abc) \Leftrightarrow ab||ac$.

Proof. By the theorems of § 1.2 together with 2.1.0′,

$$\boldsymbol{L}(abc) \overset{\mid\Rightarrow}{\underset{\Leftarrow\mid}{}} (a = b \vee a = c \vee [ab] = [ac]) \overset{\mid\Rightarrow}{\underset{\Leftarrow\mid}{}} ab||ac. \blacksquare$$

The axiom 2.1.2 is an elementary form of the Euclid postulate; it can be strengthened to a transitivity law:

Th 2.1.4. $a \neq b \wedge ab||pq \wedge ab||rs \Rightarrow pq||rs$.

Proof. Let us assume

$$a \neq b, \quad ab||pq, \quad ab||rs$$

and use 2.1.0′. We can assume additionally

$$p \neq q \quad \text{and} \quad r \neq s.$$

Let $P = [ab]$, $Q = [pq]$, $Q' = [rs]$; then

(1) $P = Q$ or $P \cap Q = \emptyset$

and

(2) $P = Q'$ or $P \cap Q' = \emptyset$.

If either $P = Q$ or $P = Q'$ then $pq||rs$ (by (2) or (1) respectively). If neither $P = Q$ nor $P = Q'$ then

$$P \cap Q = \emptyset \quad \text{and} \quad P \cap Q' = \emptyset$$

and thus, by the condition (ii) (which is a consequence of 2.1.2), $Q \cap Q' \neq \emptyset$ implies $Q = Q'$. Therefore $pq||rs$. ∎

After the discussion in § 2.1, it becomes doubtful, whether collinearity is really a proper primitive notion for affine planes; it seems that parallelity is more suitable.

Given a set S and a quaternary relation $||$ on S, the pair $(S, ||)$ is said to be a *parallelity plane* whenever the following axioms are satisfied in $(S, ||)$:

Ax 2.2.0. $ab||ba$,

Ax 2.2.1. $ab||cc$,

Ax 2.2.2. $a \neq b \wedge ab||pq \wedge ab||rs \Rightarrow pq||rs$,

Ax 2.2.3. $ab||ac \Rightarrow ba||bc$,

Ax 2.2.4. $\exists abc \sim (ab||ac)$,

Ax 2.2.5. $\forall abp\, \exists q\, (ab||pq \wedge p \neq q)$,

Ax 2.2.6. $\sim (ab||cd) \Rightarrow \exists p\, (pa||pb \wedge pc||pd)$.

(The figures 1, 2, 3 illustrate the axioms 2.2.4, 2.2.5, 2.2.6 respectively.)[6]

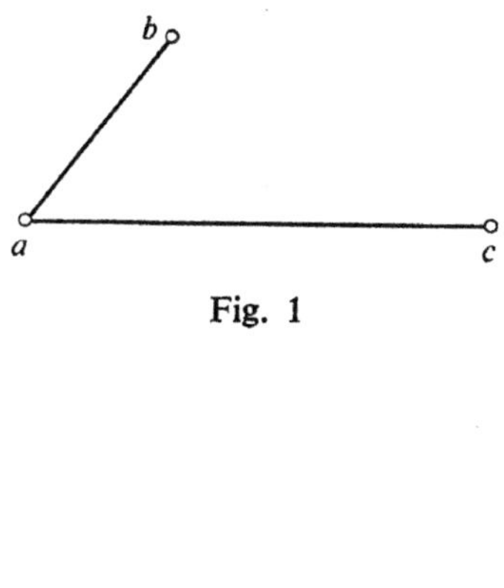

Fig. 1

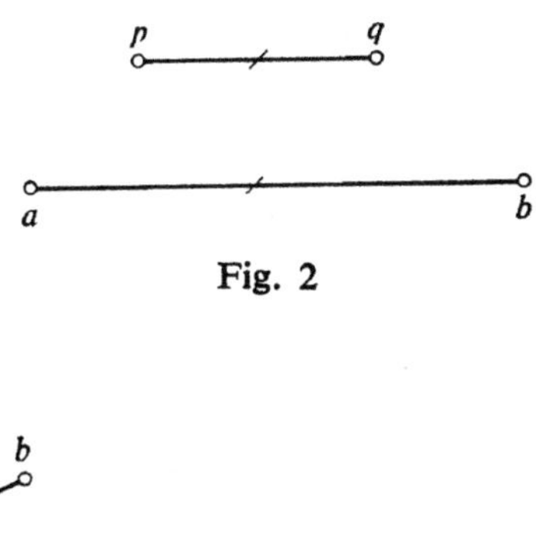

Fig. 2

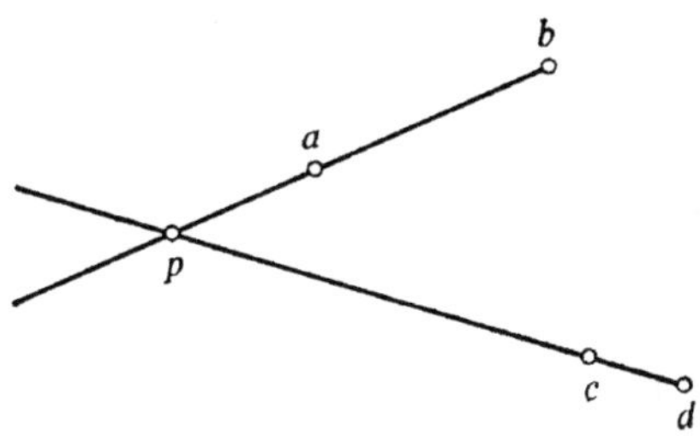

Fig. 3

[6] In figure 1 we indicate the way of marking parallel segments.

Thus, the set of sentences

$$\text{AX } 2.2 = \{2.2.0, 2.2.1, \ldots, 2.2.6\}$$

is an axiom system for the parallelity planes.

Th 2.2.7. $ab||ab$.

P r o o f. By 2.2.1, it follows that $ba||bb$; by 2.2.3

$$ba||bb \models\Rightarrow ab||ab. \blacksquare$$

Th 2.2.8. $ab||cd \Rightarrow cd||ab$.

P r o o f. By 2.2.1

$$a = b \models\Rightarrow cd||ab;$$

by 2.2.2

$$a \neq b \wedge ab||cd \wedge ab||ab \models\Rightarrow cd||ab,$$

which (by 2.2.7) completes the proof. ■

By 2.2.1 and 2.2.8 we have

Th 2.2.9. $aa||cd$.

As a consequence of 2.2.9, 2.2.0, and 2.2.2 we obtain

Th 2.2.10. $ab||cd \Rightarrow ba||cd$.

From 2.2.10 we derive (by means of 2.2.8)

Th 2.2.11. $ab||cd \Rightarrow ab||dc$.

Let us define collinearity:

Def 2.2.12. $L(abc) \Leftrightarrow ab||ac$.

By this definition

$$2.2.1 \models\Rightarrow 1.1.1',$$

$$2.2.2 \models\Rightarrow 1.1.2',$$

$$2.2.3 \models\Rightarrow 1.1.3',$$

$$2.2.4 \models\Rightarrow \dim;$$

whence

(i) $\text{AX } 2.2 \models\Rightarrow \text{AX } 1.1.$

So, we can now use the notion of line and apply all the theorems of §§ 1.1 and 1.2.

Th 2.2.13. $ab||cd \Rightarrow (a = b \vee c = d \vee [ab] = [cd]$
$\vee [ab] \cap [cd] = \emptyset)$.

P r o o f. Let us assume

(1) $ab||cd, \quad a \neq b, \quad c \neq d, \quad [ab] \cap [cd] \neq \emptyset.$

We are going to show

(2) $[ab] = [cd]$.

By the definition 2.2.12 and theorems of § 1.2, we conclude: if $a = c$, then

$$ab||cd \mapsto ab||ad \mapsto L(abd) \mapsto [ab] = [ad] = [cd].$$

Let $a \neq c$ and

(3) $p \in [ab] \cap [cd]$.

Then $p \neq a$ or $p \neq c$. By 2.2.8, we may assume without the loss of generality that $p \neq a$. In view of (3), to prove (2) it suffices to show that $a \in [cd]$. From (3) it follows that

$$ab||ap \quad \text{and} \quad cd||cp;$$

by 2.2.2

$$a \neq b \wedge ab||cd \wedge ab||ap \mapsto cd||ap.$$

Thus, if $p = c$, then $cd||ac$ and by 2.2.11,

$$cd||ac \mapsto cd||ca \mapsto L(cda) \mapsto a \in [cd].$$

If, on the contrary, $p \neq c$, then by (3), 2.2.2, 2.2.10, and 2.2.11,

$$c \neq d \wedge cd||cp \wedge cd||ap \mapsto cp||ap \mapsto pc||pa \mapsto L(pca)$$
$$\mapsto a \in [pc] = [cd]. \blacksquare$$

Th 2.2.14. $a, b, c, d \in P \Rightarrow ab||cd.$

P r o o f. Let $a, b, c, d \in P$. By 1.2.5

$$L(bad), \quad L(bac), \quad L(cbd),$$

and, by 2.2.12, 2.2.10, 2.2.8, 2.2.2

$$b = c \wedge L(bad) \Rightarrow b = c \wedge ba||bd \Rightarrow ba||cd \Rightarrow ab||cd,$$

$$b \neq c \wedge L(bac) \wedge L(cbd) \Rightarrow b \neq c \wedge ba||bc \wedge cb||cd$$

$$\Rightarrow b \neq c \wedge ab||bc \wedge bc||cd \Rightarrow ab||cd. \blacksquare$$

By 2.2.9, 2.2.1, 2.2.14, and 2.2.6, using 2.2.12 we obtain the theorem converse to 2.2.13:

Th 2.2.15. $(a = b \vee c = d \vee [ab] = [cd] \vee [ab] \cap [cd] = \emptyset) \Rightarrow ab||cd.$

Theorems 2.2.13 and 2.2.15 give the definition 2.1.0. Thus, on the base of 2.2.12,

$$\text{Ax } 2.2 \Rightarrow 2.1.0$$

(in fact the axioms 2.2.4 and 2.2.5 were not needed in the proof of 2.1.0). Moreover,

$$2.2.5 = 2.1.1$$

and, on the base of 2.2.12,

$$2.2.2 \Rightarrow 2.1.2.$$

Whence

(ii) $\quad$ AX 2.2 $\Rightarrow$ AX 2.1.

By (i) and (ii) we have

Theorem 2.2.16. *If the structure* $(S, ||)$ *is a parallelity plane and the relation* $L \subset S \times S \times S$ *is defined by* 2.2.12, *then* $(S, ||, L)$ *is an affine plane.*[7]

Conversely,

Theorem 2.2.17. *If the structure* $(S, L, ||)$ *is an affine plane, then it satisfies* 2.2.12, *and its reduct* $(S, ||)$ *is a parallelity plane.*

Proof. It suffices to notice the following inferences and identities:

$$2.1.0' \Rightarrow 2.2.0, 2.2.1,$$

$$2.1.3 \Rightarrow 2.2.3, 2.2.4,$$

$$\{2.1.0', 2.1.3\} \Rightarrow 2.2.6,$$

---

[7] We identify the structure $(S, L, ||)$ with $(S, ||, L)$.

$$2.1.4 = 2.2.2,$$

$$2.1.1 = 2.2.5,$$

$$2.1.3 = 2.2.12. \blacksquare$$

Thus the affine planes $(\boldsymbol{S}, \boldsymbol{L}, ||)$ coincide with definitional expansions of parallelity planes $(\boldsymbol{S}, ||)$ by the relation $\boldsymbol{L}$ defined by the formula 2.2.12. So, in fact, the notion of parallelity plane does not differ essentially from the notion of affine plane. But, an advantage of the first one is, that a parallelity plane can be described by simple axioms which concern only its primitive notion. Collinearity may be considered as a subrelation $\{(abcd): ab||cd \wedge a = c\}$ of parallelity, and each formula $\boldsymbol{L}(abc)$ as an abreviation of the corresponding formula $ab||cd$, obtained by the identification of the first argument with the third. Therefore, we do not distinguish between these two notions; in the sequel we shall consistently deal with the notion of parallelity plane. The class of all the parallelity planes will be denoted by the symbol $\mathscr{A}$.

We pass now to the most important affine notion. Of course, we may use all the theorems of §§ 2.1 and 2.2.

2.3. PARALLEL LINES

The §§ 2.1 and 2.2 concern implicitly the parallelity relation and its properties. We are now going to describe these properties explicitly.

Def 2.3.0. $P||Q \Leftrightarrow \exists abcd\,(P = [ab] \wedge Q = [cd] \wedge ab||cd)$.

By 2.1.0′ we derive from this definition the well known set-theoretical definition of the Euclidean parallelity of lines:

Th 2.3.1. $P||Q \Leftrightarrow (P = Q \vee P \wedge Q = \emptyset)$.

Th 2.3.2. $\sim(P||Q) \Leftrightarrow \exists a \;\; P \cap Q = \{a\}$.

Proof: by 2.3.1 and 1.2.10. ■

Th 2.3.3. $a \neq b \wedge c \neq d \wedge ab||cd \Rightarrow [ab]\,||\,[cd]$.

Proof: by 2.3.0. ■

Th 2.3.4. $[ab]\,||\,[cd] \Rightarrow ab||cd$.

Proof: by 2.3.1 and 2.1.0′ (or 2.2.14, 2.2.2, and 2.2.8). ■

Th 2.3.5. $ab||cd \Leftrightarrow \exists PQ\,(P||Q \wedge a, b \in P \wedge c, d \in Q)$.

Proof: by 2.1.1, 2.3.0, 2.2.1, and 2.3.4. ∎

Th 2.3.6. $P||P$.

Proof: by 2.3.1 (or 2.2.7). ∎

Th 2.3.7. $P||Q \Rightarrow Q||P$.

Proof: by 2.3.1 (or 2.2.8). ∎

Th 2.3.8. $P||Q \wedge P||R \Rightarrow Q||R$.

Proof: by 2.1.4 (= 2.2.2), 2.3.3, and 2.3.4. ∎

By 2.3.6 and 2.3.8, *parallelity is an equivalence relation on the set of lines*. Its equivalence classes are called *directions*.

Th 2.3.9. $\forall p\,\forall P\,\exists Q\,(p \in Q \wedge P||Q)$.

Proof: by 2.1.1 (= 2.2.5) and 2.3.3. ∎

Th 2.3.10. $p \in Q \wedge p \in Q' \wedge P||Q \wedge P||Q' \Rightarrow Q = Q'$.

Proof: by 2.3.8 and 2.3.1. ∎

By 2.3.9 and 2.3.10, there is a unique line Q passing through a given point p and parallel to a given line P: $Q = \mathbf{L}_{||P}(p)$. Let us consider the function

$$\mathbf{L}_{||P}\colon S \to \{Q\colon P||Q\},$$

which depends on the parameter P. Let $\sim (K||P)$. Restricting the domain S of $\mathbf{L}_{||P}$ to the line K, we obtain the function

$$\mathbf{L}_{||P}|K\colon K \to \{Q\colon P||Q\},$$

which, by 2.3.7, 2.3.8, and 2.3.2, is a bijection (vide fig. 4). Let now $\sim (K||P)$ and $\sim (L||P)$. The bijection

$$(\mathbf{L}_{||P}|L)^{-1}\mathbf{L}_{||P}|K\colon K \to L,$$

which is the composition of two bijections, $\mathbf{L}_{||P}|K$ and $(\mathbf{L}_{||P}|L)^{-1}$, is referred to as the *parallel projection of the line K onto the line L in direction of P*. We shall use the symbol

$$\pi\colon K \underset{||P}{\to} L$$

to denote such a projection (vide fig. 5). So much for parallel lines.

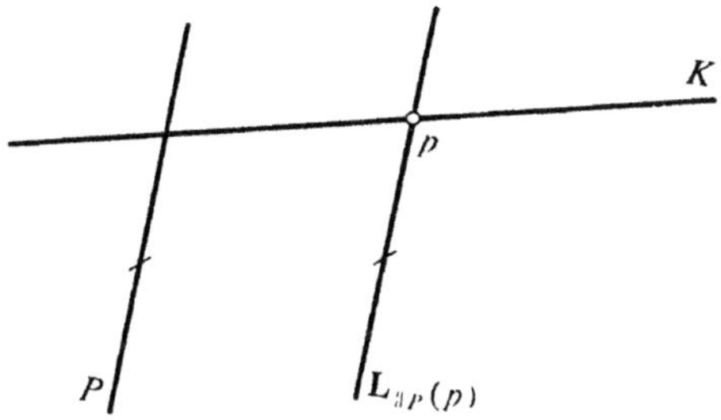

Fig. 4

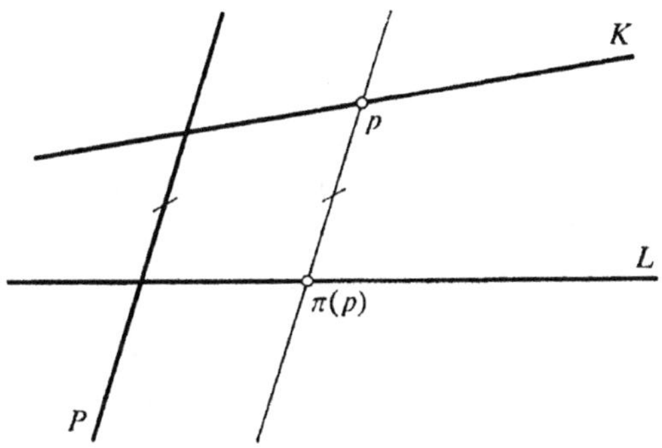

Fig. 5

Th 2.3.11. $\forall p\, \exists Q_1 Q_2 Q_3 (\neq(Q_1 Q_2 Q_3) \wedge p \in Q_1 \wedge p \in Q_2 \wedge p \in Q_3)$.

P r o o f. By the axiom dim there exist three non-collinear points a, b, c (see fig. 6). In turn, by 2.3.9, given a point p, there are lines Q_1, Q_2, Q_3 passing through p and parallel to $[ab]$, $[ac]$, and $[bc]$ respectively. By 2.3.10, the lines Q_1, Q_2, Q_3 are pairwise distinct. ■

Th 2.3.12. $\forall p\, \forall P_1 P_2\, \exists Q\, (p \in Q \wedge \sim (P_1 || Q) \wedge \sim (P_2 || Q))$.

P r o o f. By 2.3.11, given a point p there are three distinct lines Q_1, Q_2, Q_3 passing through p; by 2.3.10 at most one of them is parallel to P_1 and at most one is parallel to P_2. ■

Using the bijections defined above, we shall prove some statements concerning the number of points, $|P|$, of an arbitrary line P.

Th 2.3.13. $|K| = |L|$.[8]

---

[8] Theorem 2.3.13 is proved for an arbitrary cardinal number n, however it will be applied only in the case of natural n.

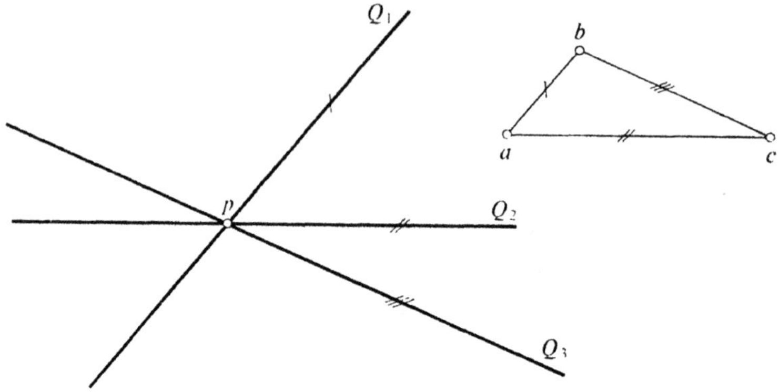

Fig. 6

Proof. By 2.3.12, there is a line P such that $\sim(P||K)$ and $\sim(P||L)$, and thus there is also a bijection $\pi\colon K \underset{||P}{\rightarrow} L$. ∎

Th 2.3.14. $|P| = |\{Q : P||Q\}|$.

Proof. By 2.3.12, there is a line K such that $\sim(K||P)$, and thus there is a bijection $\mathbf{L}_{||P}|K\colon K \rightarrow \{Q\colon P||Q\}$; moreover, by 2.3.13, $|K| = |P|$. ∎

Th 2.3.15. $|P| = n \Rightarrow |\mathbf{S}| = n^2$.

Proof. Let us notice that, by 2.3.9,

$$\mathbf{S} = \bigcup \{Q\colon P||Q\}.$$

So, it now suffices to apply 2.3.1, 2.3.13, and 2.3.14.[9] ∎

2.4. PARALLELOGRAMS AND TRAPEZOIDS

We are now going to define a quaternary *parallelogram relation* $\mathbf{P}$ on the universe $\mathbf{S}$.[10]

Def 2.4.0. $\mathbf{P}(abcd) \Leftrightarrow \sim \mathbf{L}(abc) \wedge ab||cd \wedge ac||bd.$

[9] For some informations concerning possible finite values of the power of an arbitrary line, see Pickert [75], p. 294 and Lenz [68], p. 142.

[10] The relation $\mathbf{P}$ is helpful as long as we cannot use the transitive group of translations (§ 3.1).

If $\boldsymbol{P}(abcd)$ (vide fig. 7), then the quadruple $(abcd)$ is referred to as a *parallelogram*[11], and points a, b, c, d are called its *vertices*.

Th 2.4.1. $\boldsymbol{P}(abcd) \Rightarrow \neq (abcd)$.

Proof: by 2.4.0, 1.1.1, and 2.2.12. ■

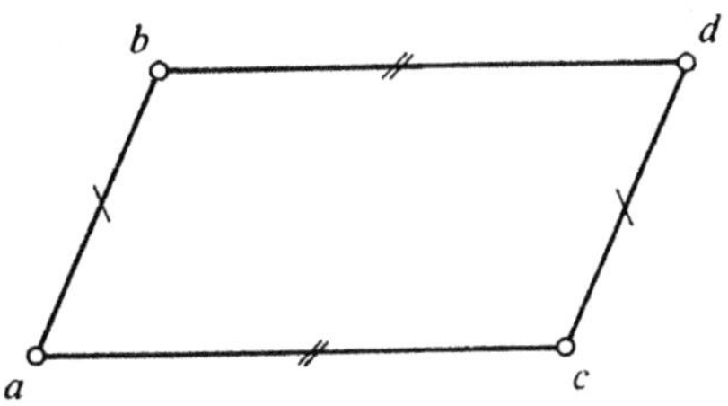

Fig. 7

Thus every two vertices of a parallelogram determine a line. The lines $[ab]$, $[cd]$, $[ac]$, $[bd]$ are referred to as *sides*, the lines $[ad]$ and $[bc]$ as *diagonals* of the parallelogram $(abcd)$.

Th 2.4.2. $\boldsymbol{P}(abcd) \Rightarrow$

$$\sim \boldsymbol{L}(abc) \wedge \sim \boldsymbol{L}(abd) \wedge \sim \boldsymbol{L}(acd) \wedge \sim \boldsymbol{L}(bcd).$$

Proof: by 2.4.0, 2.2.13, 2.4.1, and 2.2.2, 2.2.12. ■

By 2.4.0 it follows immediately

Th 2.4.3. $\boldsymbol{P}(abcd) \Rightarrow \boldsymbol{P}(acbd)$.

Th 2.4.4. $\boldsymbol{P}(abcd) \Rightarrow \boldsymbol{P}(cdab)$.

Proof: by 2.4.2, 2.2.8, 2.2.10, and 2.2.11. ■

Th 2.4.5. $\boldsymbol{P}(abcd) \Rightarrow \boldsymbol{P}(badc)$.

Proof: by 2.4.3 and 2.4.4 (or 2.4.2, 2.2.10, 2.2.11, 2.2.8).[12] ■

Let the points a, b, c satisfy the condition

$$\sim \boldsymbol{L}(abc).$$

---

[11] This term is more suitable for the quadruple $(abdc)$, but, in fact, it concerns the vectors $\overrightarrow{ab}$ and $\overrightarrow{cd}$, and therefore we prefer to use it for $(abcd)$.

[12] The proof indicated in the parentheses is more complicated, but it can be modified to a proof of the corresponding theorem on trapezoid (2.4.5′).

Then, by 2.3.4, $\sim([ab] \| [ac])$, whence $\sim(\mathbf{L}_{\|[ab]}(c) \| \mathbf{L}_{\|[ac]}(b))$ (vide fig. 8), because $\|$ is an equivalence relation. By 2.3.2, the lines $\mathbf{L}_{\|[ab]}(c)$ and $\mathbf{L}_{\|[ac]}(b)$ have a point of intersection, d; by 2.3.4, 2.4.0–2.4.2, and 2.3.3

$$\mathbf{L}_{\|[ab]}(c) \cap \mathbf{L}_{\|[ac]}(b) = \{d\} \Leftrightarrow ab\|cd \wedge ac\|bd.$$

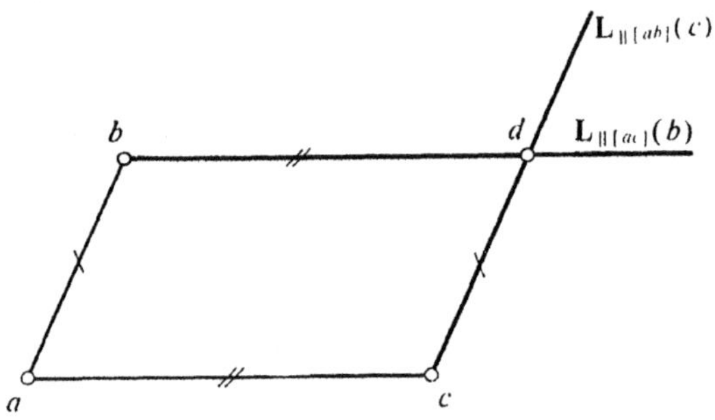

Fig. 8

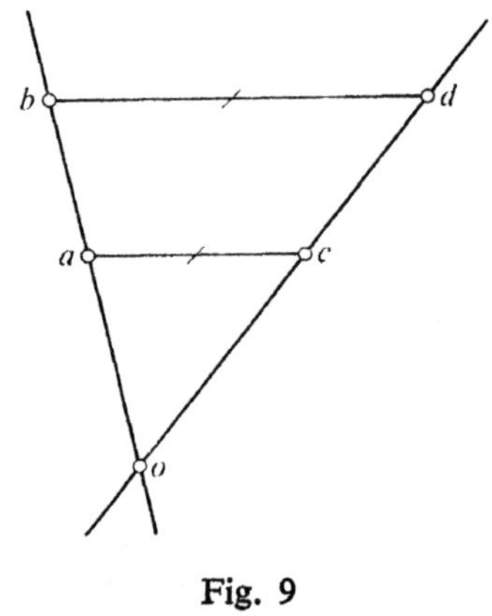

Fig. 9

In conclusion, an arbitrary triple of non-collinear points (abc) can be—in the unique way—completed to the parallelogram $(abcd)$. This is expressed by the following two statements:

Th 2.4.6. $\sim L(abc) \Rightarrow \exists d\ P(abcd).$

Th 2.4.7. $P(abcd) \wedge P(abcd') \Rightarrow d = d'.$

In view of 2.4.6 and 2.4.7, the relation P may be considered as a function of the first three arguments:

Def 2.4.8. $P(abc) = d \Leftrightarrow P(abcd).$

Th 2.4.9. $P(abc) = P(acb).$

Proof: by 2.4.8 and 2.4.3. ∎

As an analogue of the parallelogram relation $\boldsymbol{P}$, we introduce the *trapezoid relation*, $\boldsymbol{T}$.[13] It is convenient to treat $\boldsymbol{T}$ as a 5-ary relation, connecting the four *vertices* a, b, c, d of a trapezoid with the point of intersection o of its non-parallel sides $[ab]$ and $[cd]$ (vide fig. 9). In the applications this point o will be always fixed. Thus we treat o as a parameter, and pass from the relation $\boldsymbol{T}$ to $\boldsymbol{T}_o$:

Def 2.4.0′. $\boldsymbol{T}_o(abcd) \Leftrightarrow$

$$o \neq a, b, c, d \wedge \sim \boldsymbol{L}(abc) \wedge \boldsymbol{L}(oab) \wedge \boldsymbol{L}(ocd) \wedge ac || bd.$$

The properties of $\boldsymbol{P}$ formulated in 2.4.1, 2.4.2 and 2.4.4–2.4.7 (but not in 2.4.3) transfer without any change on $\boldsymbol{T}_o$:

Th 2.4.1′. $\boldsymbol{T}_o(abcd) \Rightarrow \neq (abcd)$.

Th 2.4.2′. $\boldsymbol{T}_o(abcd) \Rightarrow$

$$\sim \boldsymbol{L}(abc) \wedge \sim \boldsymbol{L}(abd) \wedge \sim \boldsymbol{L}(acd) \wedge \sim \boldsymbol{L}(bcd).$$

Th 2.4.4′. $\boldsymbol{T}_o(abcd) \Rightarrow \boldsymbol{T}_o(cdab)$.

Th 2.4.5′. $\boldsymbol{T}_o(abcd) \Rightarrow \boldsymbol{T}_o(badc)$.

Th 2.4.6′. $o \neq a, b, c \wedge \sim \boldsymbol{L}(abc) \wedge \boldsymbol{L}(oab) \Rightarrow \exists d\ \boldsymbol{T}_o(abcd)$.

Th 2.4.7′. $\boldsymbol{T}_o(abcd) \wedge \boldsymbol{T}_o(abcd') \Rightarrow d = d'$.

We leave to the reader the simple proofs of these theorems; they are quite similar to the proofs of corresponding statements on parallelogram. In view of 2.4.6′ and 2.4.7′, the relation $\boldsymbol{T}_o$ can be considered as a function of the first three arguments:

Def 2.4.8′. $\boldsymbol{T}_o(abc) = d \Leftrightarrow \boldsymbol{T}_o(abcd)$.

2.5. TRANSLATIONS AND HOMOTHETIES

Let us consider a parallelity plane $\mathfrak{A} = (\boldsymbol{S}, ||)$. The group of its automorphisms, Aut, consists of the bijections $\varphi: \boldsymbol{S} \to \boldsymbol{S}$ which preserve the relation $||$, i.e. satisfy the condition

(i) $pq || rs \Leftrightarrow p^\varphi q^\varphi || r^\varphi s^\varphi$.

The identity bijection will be denoted by the letter $\boldsymbol{I}$.

---

[13] The relation $\boldsymbol{T}$ will be helpful as long as we cannot use the transitive groups of homotheties (§ 3.1).

The condition

(ii) $pq \| p^{\varphi}q^{\varphi}$

is stronger than (i); it distinguishes among all the automorphisms the so called *dilatations*. They constitute a subgroup Dil of Aut, moreover a normal subgroup (see 2.5.4 (i)). The Greek letter δ will be always used to denote a dilatation. Let us rewrite the condition (ii) as

Th 2.5.0. $pq \| p^{\delta}q^{\delta}$.

We are now going to investigate the possible number of fixed points of a dilatation different from the identity.

Th 2.5.1. $a^{\delta} = a \Rightarrow a \in [pp^{\delta}]$.

Proof: by 2.5.0 and 2.1.3. ■

Lem 2.5.2. $\sim L(abp) \wedge a^{\delta} = a \wedge b^{\delta} = b \Rightarrow p^{\delta} = p$.[14]

Proof: by 2.5.1. ■

Th 2.5.3. $a \neq b \wedge a^{\delta} = a \wedge b^{\delta} = b \Rightarrow \delta = I$.

Proof. By 1.1.5 and 2.5.2, any pair of distinct fixed points a and b of a dilatation δ can be extended to a triple of non-collinear fixed points a, b, c of δ:

$$\sim L(abc) \wedge a^{\delta} = a \wedge b^{\delta} = b \wedge c^{\delta} = c.$$

For an arbitrary point $p \in \boldsymbol{S}$, one of the following conditions is satisfied:

$$\sim L(abp) \quad \text{or} \quad \sim L(bcp) \quad \text{or} \quad \sim L(cap);$$

thus, by 2.5.2, p is a fixed point of δ. Hence $\delta = I$. ■

In conclusion, *every dilatation different from I has at most one fixed point*. If δ has no fixed points, then it is called a *translation*; if δ has one fixed point o, then it is called a *homothety with center o*. The identity dilatation is considered as both a translation and a homothety with any point as center. The set of all the translations is denoted by the symbol Tr; as variables for its elements we use the Greek letters α, β, γ. The set of the homotheties with center o is denoted by the symbol

[14] The statement 2.5.2 is treated as a lemma, because it is a particular case of the next theorem 2.5.3, and is needed (twice) only to its proof. Further lemmas are of the same character.

Ht_o; the set of all the homotheties by the symbol Ht. As variables for homotheties we use the Greek letters $\varkappa$, λ.

Th 2.5.4. *If* $\alpha \in \mathsf{Tr}$, $\varkappa \in \mathsf{Ht}_o$, $\delta \in \mathsf{Dil}$ *and* $\varphi \in \mathsf{Aut}$, *then*

(i) $\varphi^{-1}\delta\varphi \in \mathsf{Dil}$

(ii) $\varphi^{-1}\varkappa\varphi \in \mathsf{Ht}_{o^\varphi}$

(iii) $\varphi^{-1}\alpha\varphi \in \mathsf{Tr}$.[15]

Proof. (i) Let $\delta \in \mathsf{Dil}$. Then

$$p^{\varphi^{-1}}q^{\varphi^{-1}} \| p^{\varphi^{-1}\delta}q^{\varphi^{-1}\delta},$$

whence

$$pq \| p^{\varphi^{-1}\delta\varphi}q^{\varphi^{-1}\delta\varphi};$$

thus $\varphi^{-1}\delta\varphi \in \mathsf{Dil}$.

(ii) Let $\varkappa \in \mathsf{Ht}_o$, i.e. $\varkappa \in \mathsf{Dil}$ and $o^\varkappa = o$. Then, by (i), $\varphi^{-1}\varkappa\varphi \in \mathsf{Dil}$, and

$$(o^\varphi)^{\varphi^{-1}\varkappa\varphi} = o^{\varkappa\varphi} = o^\varphi,$$

whence $\varphi^{-1}\varkappa\varphi \in \mathsf{Ht}_{o^\varphi}$.

(iii) Let $\alpha \in \mathsf{Tr}$. By (i), $\varphi^{-1}\alpha\varphi \in \mathsf{Dil}$. For an arbitrary $p \in S$ $p^{\varphi^{-1}\alpha\varphi} = p \Rightarrow p^{\varphi^{-1}\alpha} = p^{\varphi^{-1}} \Rightarrow \alpha = I \Rightarrow \varphi^{-1}\alpha\varphi = I$; thus $\varphi^{-1}\alpha\varphi \in \mathsf{Tr}$. ∎

Th 2.5.5. $p \in [aa^\delta] \Rightarrow p^\delta \in [aa^\delta]$.

Proof. Let $p \in [aa^\delta]$. Then

(1) $a^\delta \in [ap] \cap [a^\delta p^\delta]$;

if, moreover, $p \neq a$, then by 2.5.0,

(2) $[ap] \| [a^\delta p^\delta]$.

From (1) and (2) it follows that $[ap] = [a^\delta p^\delta]$ for $p \neq a$. Therefore in any case $p^\delta \in [aa^\delta]$. ∎

Th 2.5.6. $aa^\alpha \| bb^\alpha$.

Proof. Theorem is obvious for $\alpha = I$. Let $\alpha \neq I$. Then $a \neq a^\alpha$ and $b \neq b^\alpha$. By 2.5.5, if the lines $[aa^\alpha]$ and $[bb^\alpha]$ have a point of intersection p, then they have two distinct common points p and p^α, and thus they coincide. Hence $[aa^\alpha] \| [bb^\alpha]$. ∎

---

[15] In this section and in § 3 and § 4, the composition of maps φ_1 and φ_2 is denoted by "$\varphi_1\varphi_2$" instead of "$\varphi_2\varphi_1$" (comp. § 2.3, p. 28 or § 7.5, p. 150). Such a notation is justified by the exponential form used here for the value of a function (M.M.)

In view of 2.5.6, for $\alpha \neq I$ all the lines $[pp^{\alpha}]$ constitute one direction; it is referred to as *direction of* α.[16] Consider a line P. The subset of Tr consisting of I and of all the translations which have the direction of P will be denoted by the symbol Tr_P (vide fig. 10). So, we have

Th 2.5.7. $\alpha \in \mathrm{Tr}_P \Leftrightarrow \forall p\ [pp^{\alpha}] \| P$ (for $\alpha \neq I$).

Using 2.5.7 one easily obtains the following two theorems:

Th 2.5.8. (i) $P \| Q \Leftrightarrow \mathrm{Tr}_P = \mathrm{Tr}_Q$.

(ii) $\sim (P \| Q) \Leftrightarrow \mathrm{Tr}_P \cap \mathrm{Tr}_Q = \{I\}$.

Th 2.5.9. (i) $a \in P \wedge \alpha \in \mathrm{Tr}_P \Rightarrow a^{\alpha} \in P$.

(ii) $a \in P \wedge a^{\alpha} \in P \Rightarrow \alpha \in \mathrm{Tr}_P$.

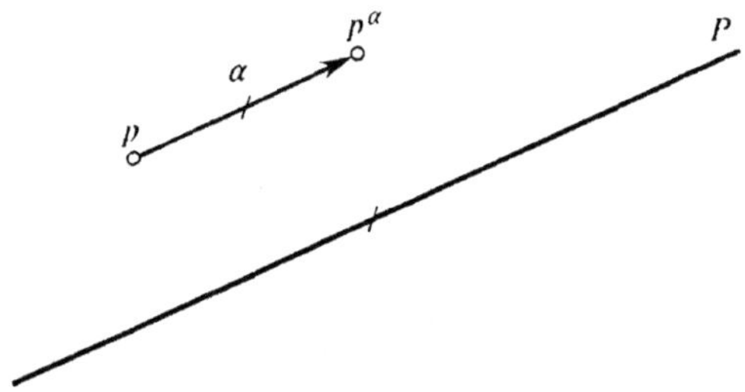

Fig. 10

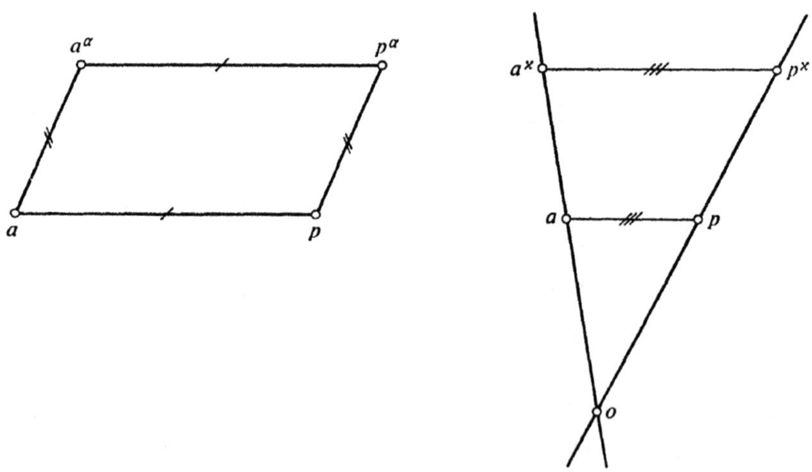

Fig. 11

[16] This direction is the *improper center of* α; it will always be represented by one of its lines.

Let us fix two points a and b and consider the (exponential[17]) equation $a^\delta = b$ with the unknown δ. We are going to check, how many solutions it may have in the set Tr and in the set Ht_o for $o \neq a$.

Th 2.5.10. (i) $\sim L(aa^\alpha p) \Rightarrow P(aa^\alpha pp^\alpha)$.

(ii) $o^\varkappa = o \wedge o \neq a, p \wedge \sim L(aa^\varkappa p) \Rightarrow T_o(aa^\varkappa pp^\varkappa)$.

Proof: (i) by 2.5.6 and 2.5.0, (ii) by 2.5.1 and 2.5.0. ∎

The figure 11 illustrates theorem 2.5.10.

Lem 2.5.11. (i) $a^\alpha = a^\beta \wedge p \notin [aa^\alpha] \Rightarrow p^\alpha = p^\beta$.

(ii) $o^\varkappa = o^\lambda = o \wedge o \neq a, p \wedge a^\varkappa = a^\lambda \wedge p \notin [aa^\varkappa] \Rightarrow p^\varkappa = p^\lambda$.

Proof. (i): If $a^\alpha = a^\beta$ and $p \notin [aa^\alpha]$, then by 2.5.10(i)

$$p^\alpha = P(aa^\alpha p) = P(aa^\beta p) = p^\beta.$$

Proof of (ii) is analogous to the proof of (i); it is based on 2.5.10(ii). ∎

Th 2.5.12. (i) $a^\alpha = a^\beta \Rightarrow \alpha = \beta$.

(ii) $o^\varkappa = o^\lambda = o \wedge o \neq a \wedge a^\varkappa = a^\lambda \Rightarrow \varkappa = \lambda$.

Proof. (i): If $a^\alpha = a^\beta = a$, then $\alpha = \beta = I$. Let

(3) $a^\alpha = a^\beta \neq a$.

Thus $\alpha, \beta \neq I$. Let us fix an arbitrary point $b \notin [aa^\alpha] = [aa^\beta]$. By 2.5.11(i)

(4) $b^\alpha = b^\beta$;

moreover

$$[aa^\alpha] \neq [bb^\alpha],$$

and, by 2.5.6,

$$[aa^\alpha] \parallel [bb^\alpha].$$

Thus, an arbitrary point $p \in S$ does not lie on at least one of the lines $[aa^\alpha]$ and $[bb^\alpha]$; hence, in view of (3) and (4), by 2.5.11 (i) we get $p^\alpha = p^\beta$. Therefore $\alpha = \beta$.

The proof of (ii) is similar to the proof of (i). It is based on 2.5.11 (ii) and 2.5.1. ∎

---

[17] Since the image of the point a by the mapping δ is denoted by the symbol a^δ, we refer to the equation $a^\delta = b$ as an *exponential* one, to emphasize that δ is the unknown.

By 2.5.12, *the exponential equation* $a^\delta = b$ *has at most one solution in the set* Tr *and at most one solution in each set* Ht_o *for* $o \neq a$. It enables us to proof

Th 2.5.13. *Each of the sets* Tr *and* Ht_o *determines a subgroup of the group* Dil.

P r o o f. If $\alpha \in \mathsf{Tr}$, then for arbitrary $a \in \boldsymbol{S}$

$$a^{\alpha^{-1}} = a \mathrel{|\!\!\Rightarrow} a = a^\alpha \mathrel{|\!\!\Rightarrow} \alpha = \boldsymbol{I} \mathrel{|\!\!\Rightarrow} \alpha^{-1} = \boldsymbol{I}$$

and thus $\alpha^{-1} \in \mathsf{Tr}$. If $\alpha, \beta \in \mathsf{Tr}$, then by 2.5.12 (i), for arbitrary $a \in \boldsymbol{S}$

$$a^{\alpha\beta} = a \mathrel{|\!\!\Rightarrow} a^\alpha = a^{\beta^{-1}} \mathrel{|\!\!\Rightarrow} \alpha = \beta^{-1} \mathrel{|\!\!\Rightarrow} \alpha\beta = \boldsymbol{I}$$

and thus $\alpha\beta \in \mathsf{Tr}$. For any set Ht_o the similar argumentation is based upon 2.5.12 (ii). ■

Let us notice that, by 2.5.4 (iii), the group Tr is even a normal subgroup of Dil.

The composition of traslations will be denoted by "+"; the composition of homotheties with center o by "·". Under this notation the group of translations coincides with the structure $(\mathsf{Tr}, \boldsymbol{I}, +)$, and the group of homotheties with center o coincides with $(\mathsf{Ht}_o, \boldsymbol{I}, \cdot)$. Using 2.5.7 we easily obtain

Th 2.5.14. *For an arbitrary line* P, *the set* Tr_P *determines the subgroup* $(\mathsf{Tr}_P, \boldsymbol{I}, +)$ *of the group* $(\mathsf{Tr}, \boldsymbol{I}, +)$.[18]

We shall write either *the set* Tr or *the group* Tr, to distinguish the universe of the group from the group itself. The same concerns the symbols "Tr_P" and "Ht_o".[19]

We shall now deal with the group Tr. The following two lemmas lead to a very weak but already sufficient condition for the group Tr to be commutative.

Lem 2.5.15. $\sim L(aa^\alpha a^\beta) \Rightarrow \alpha+\beta = \beta+\alpha$.

P r o o f. By 1.1.3, 2.5.10 (i), 2.4.9, 2.5.12 (i)

$$\begin{aligned} &\sim L(aa^\alpha a^\beta) \mathrel{|\!\!\Rightarrow} \sim L(aa^\beta a^\alpha) \wedge \sim L(aa^\alpha a^\beta) \\ &\mathrel{|\!\!\Rightarrow} a^{\alpha+\beta} = \boldsymbol{P}(aa^\beta a^\alpha) = \boldsymbol{P}(aa^\alpha a^\beta) = a^{\beta+\alpha} \mathrel{|\!\!\Rightarrow} \alpha+\beta = \beta+\alpha. \ ■ \end{aligned}$$

---

[18] Such a notation will be justified in § 3.4, where the groups $(\mathsf{Tr}_P, I, +)$ and $(\mathsf{Ht}_o, I, \cdot)$ will be mapped isomorphically onto the additive group and the multiplicative group of a certain field of points of the line $P \ni o$.

[19] In general, in the case of transformation groups, the symbol denoting the universe of a group will also be used to denote the group itself.

Lem 2.5.16. $\sim L(aa^\alpha a^\beta) \Rightarrow \alpha+\gamma = \gamma+\alpha$.

Proof. We may assume $\gamma \neq I$. Let

(1) $\sim L(aa^\alpha a^\gamma)$.

In view of 2.5.15, it suffices to consider the case

(2) $L(aa^\alpha a^\gamma)$.

Then

(3) $\sim L(aa^\gamma a^\beta)$.

From (1) and (3), by 2.5.15, it follows that

(4) $\alpha+\beta = \beta+\alpha$

and $\gamma+\beta = \beta+\gamma$, which implies

(5) $-\beta+\gamma = \gamma-\beta$.

From (1) and (4) by 2.5.10 (i) and 2.4.2 it follows that $\sim L(aa^\alpha a^{\alpha+\beta})$, whence, by (2), $\sim L(aa^{\alpha+\beta}a^\gamma)$ as well. Thus, by 2.5.15 we obtain

(6) $\alpha+\beta+\gamma = \gamma+\alpha+\beta$.

Finally, by (5) and (6)

$$\alpha+\gamma = \alpha+\beta-\beta+\gamma = \alpha+\beta+\gamma-\beta = \gamma+\alpha+\beta-\beta = \gamma+\alpha. \blacksquare$$

Th 2.5.17. $\sim L(aa^\alpha a^\beta) \Rightarrow \gamma+\gamma' = \gamma'+\gamma$.

Proof. We may assume $\gamma \neq I$. Then $a \neq a^\gamma$, which together with $\sim L(aa^\alpha a^\beta)$ implies

$$\sim L(aa^\gamma a^\alpha) \quad \text{or} \quad \sim L(aa^\gamma a^\beta).$$

By 2.5.16, in both cases $\gamma+\gamma' = \gamma'+\gamma$. ∎

According to 2.5.17, the group Tr is commutative, if there exist two translations α and β such that $\sim L(aa^\alpha a^\beta)$ for some point $a \in S$; in other words, Tr is commutative if it does not coincide with any of its subgroups Tr_P. Hence, we have

Th 2.5.18. *If* Tr $\neq$ Tr_P *for every P, then the group* Tr *is commutative.*

From 2.5.18 we obtain immediately

Th 2.5.19. *The group* Tr *is commutative if and only if each of its subgroups* Tr_P *is commutative.*

The problem of commutativity of a group Ht_o will be discussed in § 3.

Using 2.5.18 we obtain the following two theorems:

Th 2.5.20. $p^\alpha p^\beta || p^{-\alpha} p^{-\beta}$.

Proof. For $\alpha = \beta$ the theorem is obvious. For $\alpha = \boldsymbol{I}$ or $\beta = \boldsymbol{I}$ it follows by 2.5.0. So, let $\alpha, \beta \neq \boldsymbol{I}$ and $\alpha \neq \beta$. If $\mathsf{Tr} = \mathsf{Tr}_P$ for some P, then by 2.5.7 for arbitrary $p \in \boldsymbol{S}$

$$[pp^\alpha] || P \wedge [pp^\beta] || P,$$

thus

$$\boldsymbol{L}(pp^\alpha p^\beta),$$

whence

(1) $\quad [p^\alpha p^\beta] || P;$

in similar way we obtain

(2) $\quad [p^{-\alpha} p^{-\beta}] || P.$

In view of 2.3.4, from (1) and (2) it follows that $p^\alpha p^\beta || p^{-\alpha} p^{-\beta}$. If $\mathsf{Tr} \neq \mathsf{Tr}_P$ for every P, then, by 2.5.18, the group Tr is commutative. Then it suffices to substitute in 2.5.0 $p^{-\alpha}, p^{-\beta}, \alpha+\beta$ for p, q, δ. ■

Figure 12 illustrates theorem 2.5.20.

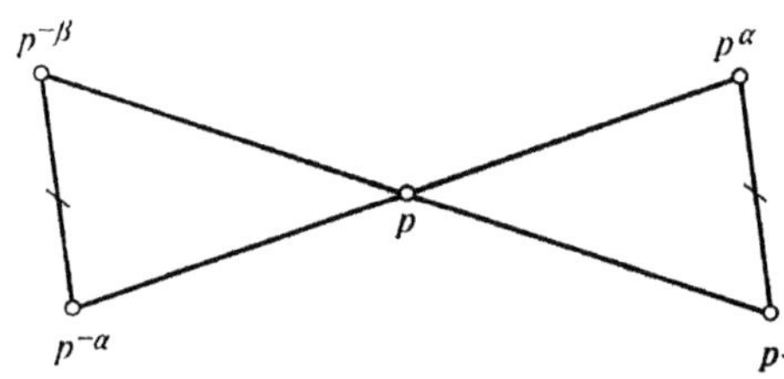

Fig. 12

Th 2.5.21. $p^\alpha p^\beta || q^\alpha q^\beta$.

Proof. For $\alpha = \beta$ the theorem is obvious. For $\alpha = \boldsymbol{I}$ or $\beta = \boldsymbol{I}$ it follows by 2.5.6. Let $\alpha, \beta \neq \boldsymbol{I}$, $\alpha \neq \beta$. If $\mathsf{Tr} = \mathsf{Tr}_P$ for some P, then the argumentation is analogous to that used in the proof of 2.5.20.

If $\mathrm{Tr} \neq \mathrm{Tr}_P$ for every P, then we substitute in 2.5.6 p^α, q^α, $-\alpha+\beta$ for a, b, α. ■

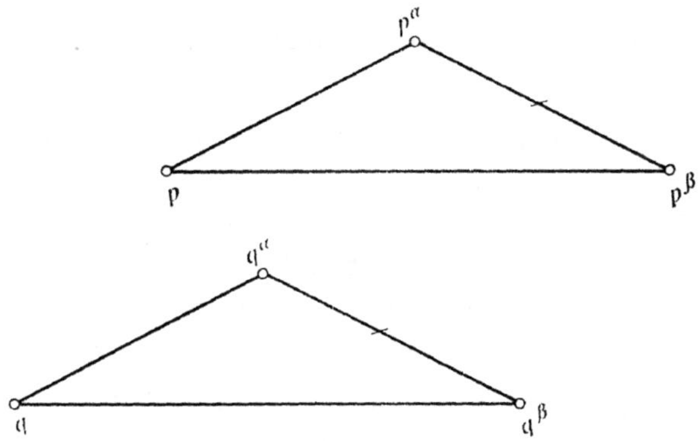

Fig. 13

Figure 13 illustrates theorem 2.5.21.

CHAPTER 3

DESARGUEAN AND PAPPIAN AFFINE PLANES

3.1. TRANSITIVE GROUPS OF TRANSLATIONS AND HOMOTHETIES

Let us consider an arbitrary parallelity plane $\mathfrak{A} = (\boldsymbol{S}, \|)$, its set of dilatations Dil, set of translations Tr, and the set of homotheties Ht. Given any set of points $S \subseteq \boldsymbol{S}$ and any set of dilatations $D \subseteq$ Dil, the set D is said to be *transitive on* S, if for every two points $a, b \in S$ the exponential equation $a^{\delta} = b$ has a solution in D. In applications, the set D consists of bijections $\delta: S \to S$ and constitutes a group of transformations.

If the set of translations, Tr is transitive on the whole universe $\boldsymbol{S}$, then we simply say that it is *transitive*, or, that the *group* Tr *is transitive*. Then the parallelity plane $\mathfrak{A}$ is referred to as a *translation plane*. Hence, $\mathfrak{A}$ is a translation plane, if the following condition is satisfied in $\mathfrak{A}$:

(i) $\forall ab \exists \alpha \; a^{\alpha} = b.$

Analogically, if the set Tr_P is transitive on the line P, then we simply say that it is *transitive*, or that the *group* Tr_P *is transitive*. Evidently, the transitivity of the set Tr is equivalent to the transitivity of all the sets Tr_P.

Let us fix a point $o \in \boldsymbol{S}$. If for every line $P \ni o$ the set Ht_o is transitive on $P - \{o\}$, then we simply say that it is *transitive*, or, that the *group* Ht_o *is transitive*. If the set Ht_o is transitive for every point $o \in \boldsymbol{S}$, then Ht is said to be *transitive* and the plane $\mathfrak{A}$ is referred to as a *homothety plane*.[1] Hence, $\mathfrak{A}$ is a homothety plane if the following condition is satisfied in $\mathfrak{A}$:

(ii) $o \neq a, b \wedge \boldsymbol{L}(oab) \Rightarrow \exists \varkappa \, (o^{\varkappa} = o \wedge a^{\varkappa} = b).$

Th 3.1.0. *If every line consists of two points, then the group* Tr *is transitive.*

[1] For an affine plane with the transitive set of translations, the term *tsanslation plane* is in common use. As an analogue, the term *homothety plane* is introduced here for an affine plane with the transitive set of homotheties.

P r o o f. Take arbitrary two distinct points $a, b \in S$. By 2.3.15, the whole universe consists of four points: $S = \{a, b, c, d\}$. It is easy to check that the involution which permutes each of two pairs (ab) and (cd) is a translation carrying a into b. ■

Th 3.1.1. *If the group* Tr *is transitive, then it is commutative.*

P r o o f. Let us take three non-collinear points a, b, c. If the group Tr is transitive, then there exist two translations α and β such that $a^\alpha = b$ and $a^\beta = c$. Then $\sim L(aa^\alpha a^\beta)$, whence, by 2.5.17, the group Tr is transitive. ■

Th 3.1.2. *If the set* Ht *is transitive, then the group* Tr *is transitive as well.*

P r o o f. Let us assume that Ht_o is transitive for every $o \in S$. Take two distinct points $a, b \in S$. By 3.1.0, we may assume that there is a point $p \in [ab]$, different from both a and b. Then there exists a homothety $\varkappa \in \mathrm{Ht}_p$ satisfying the condition

(1) $\qquad a^\varkappa = b.$

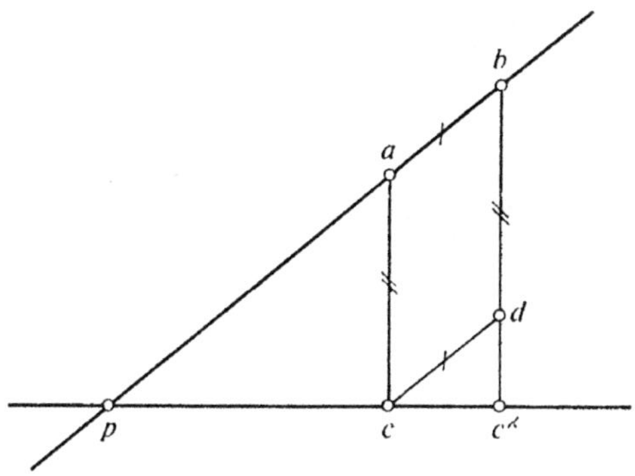

Fig. 14

Let us choose a point c such that $\sim L(abc)$, and complete the triple (abc) to the parallelogram $(abcd)$ (vide fig. 14). It is easy to see that $L(bc^\varkappa d)$ and $b \neq c^\varkappa, d$. Thus, there is a homothety $\lambda \in \mathrm{Ht}_b$ satisfying the condition

(2) $\qquad c^{\varkappa\lambda} = d;$

since $b^\lambda = b$, from (1) it follows that

(3) $\qquad a^{\varkappa\lambda} = b.$

Of course $\varkappa\lambda \in \mathsf{Dil}$. If $\varkappa\lambda$ had any fixed point, then, by 2.5.1, in view of (3) and (2), this point would lie on both lines $[ab]$ and $[cd]$. But this is impossible, because $\boldsymbol{P}(abcd)$ implies $[ab] \cap [cd] = \emptyset$. Therefore, the dilatation $\varkappa\lambda$ is a translation, and, by (3), the exponential equation $a^{\delta} = b$ has a solution in the set Tr. ∎

According to 3.1.2, every homothety plane is also a translation plane.

3.2. DESARGUES LAWS

In § 3.1 we distinguished the translation and the homothety planes. The translation planes are characterized by the condition (i), the homothety planes—by (ii). Both conditions are non-elementary, since they are expressed in terms of transformations. However, the Desargues laws, *minor* des and *major* Des[2] (see fig. 15 and 16), are elementary; as will turn out, they are equivalent to (i) and (ii) respectively, and thus will

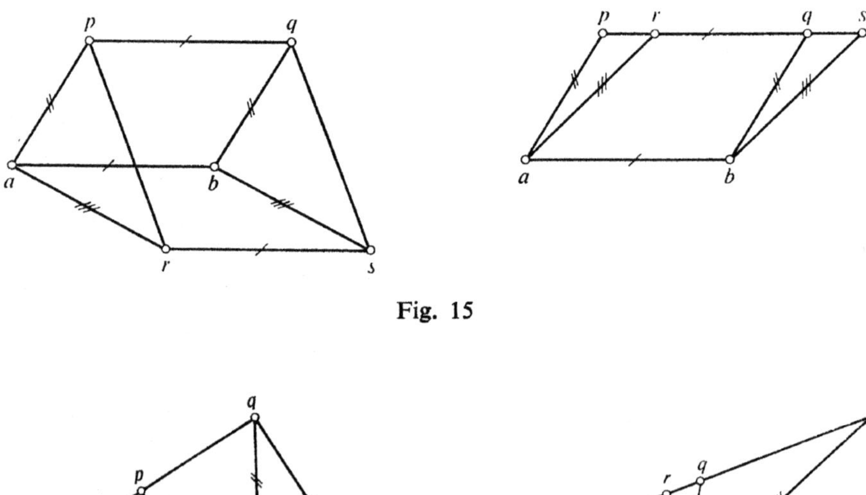

Fig. 15

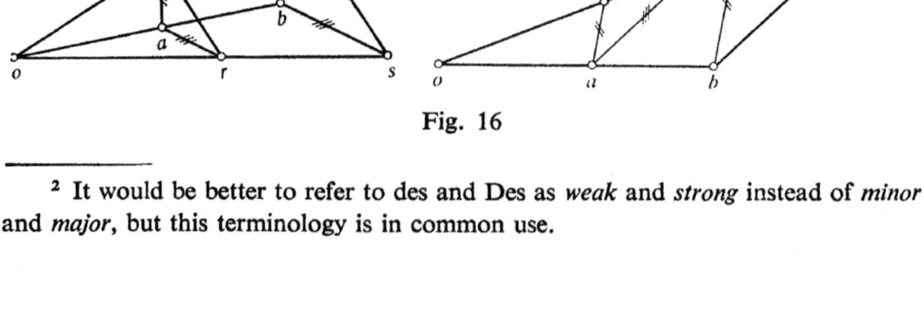

Fig. 16

[2] It would be better to refer to des and Des as *weak* and *strong* instead of *minor* and *major*, but this terminology is in common use.

be taken as axioms:

Ax des. $\sim L(abp) \wedge \sim L(abr) \wedge ab||pq \wedge ab||rs \wedge ap||bq \wedge ar||bs$
$\Rightarrow pr||qs$

Ax Des. $\sim L(abp) \wedge \sim L(abr) \wedge o \neq a, b, p, q, r, s$
$\wedge L(oab) \wedge L(opq) \wedge L(ors) \wedge ap||bq \wedge ar||bs \Rightarrow pr||qs.$

The axiom des can be expressed in terms of the parallelogram relation as the law

des′. $P(abpq) \wedge P(abrs) \Rightarrow pr||qs.$

Adjoining the assumption $\sim L(pqr)$, which excludes the trivial case $[pr] = [qs]$,[3] we get the *transitivity law for the parallelogram relation*:

des″. $P(abpq) \wedge P(abrs) \wedge \sim L(pqr) \Rightarrow P(pqrs).$

Similarly, the axiom Des can be expressed in terms of the trapezoid relation as the law

Des′. $T_o(abpq) \wedge T_o(abrs) \Rightarrow pr||qs.$

Adjoining the assumption $\sim L(pqr)$, which again excludes the trivial case $[pq] = [qs]$, we get the *transitivity law for the trapezoid relation*:

Des″. $T_o(abpq) \wedge T_o(abrs) \wedge \sim L(pqr) \Rightarrow T_o(pqrs).$

Theorem 3.2.1. *Let* $\mathfrak{A} = (S, ||)$ *be a parallelity plane.*
(i) *The set* Tr *of* $\mathfrak{A}$ *is transitive if and only if the law* des *is satisfied in* $\mathfrak{A}$.
(ii) *The set* Ht *of* $\mathfrak{A}$ *is transitive if and only if the law* Des *is satisfied in* $\mathfrak{A}$.

P r o o f. (i). Let us assume that the group Tr of $\mathfrak{A}$ is transitive, and let the points $a, b, p, q, r, s \in S$ satisfy the condition:

$$P(abpq) \quad \text{and} \quad P(abrs).$$

Then, there is a translation α such that $a^\alpha = b$, and simultaneously, by 2.5.10 (i), $p^\alpha = q$ and $r^\alpha = s$. Thus $pr||qs$ and consequently des is satisfied. Assume now that des is satisfied in $\mathfrak{A}$, and deduce the transitivity of Tr. By 3.1.0 we may additionally assume that each line consists of at least three points. Let us take two arbitrary distinct points $a_1, b_1 \in S$.

---

[3] According to our general principle (see footnote 3, p. 17), we omitted the non-essential condition $\sim L(pqr)$ in the assumptions of the axioms des and Des.

We have to construct the translation carrying a_1 onto b_1. Let

$$P_1 = [a_1 b_1].$$

By 2.3.14, there are two distinct lines P_2 and P_3 different from P_1 and parallel to P_1:

$$\neq (P_1 P_2 P_3) \quad \text{and} \quad P_1 || P_2 || P_3$$

(see fig. 17). Choose any two points $a_2 \in P_2$ and $a_3 \in P_3$, and let

$$\boldsymbol{P}(a_1 b_1 a_i) = b_i \quad \text{for} \quad i = 2, 3.$$

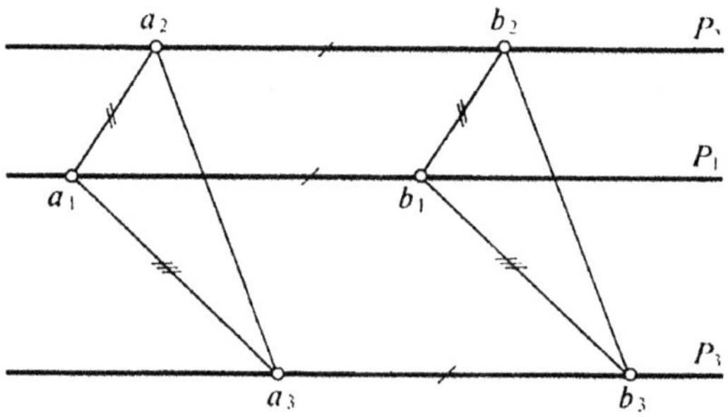

Fig. 17

Then

$$P_2 = [a_2 b_2], \quad P_3 = [a_3 b_3]$$

and

$$\boldsymbol{P}(a_1 b_1 a_2 b_2) \quad \text{and} \quad \boldsymbol{P}(a_1 b_1 a_3 b_3),$$

whence by des″

$$\boldsymbol{P}(a_2 b_2 a_3 b_3).$$

Using once more des″, we observe that

$$p \notin P_i, P_j \Rightarrow \boldsymbol{P}(a_i b_i p) = \boldsymbol{P}(a_j b_j p) \quad \text{for} \quad i, j = 1, 2, 3.$$

Thus three compatible conditions

$$p \notin P_i \Rightarrow \boldsymbol{P}(a_i b_i p) = p^{\varphi} \quad \text{for} \quad i = 1, 2, 3$$

determine a transformation φ: $\boldsymbol{S} \to \boldsymbol{S}$ (without fixed points). The map φ is a bijection, because by 2.4.2 and 2.4.5

$$p \notin P_i \wedge \boldsymbol{P}(a_i b_i p) = p' \Leftrightarrow p' \notin P_i \wedge \boldsymbol{P}(b_i a_i p') = p.$$

The bijection φ is a dilatation. Indeed, for any two points p and q there is an index $i \in \{1, 2, 3\}$ such that $p, q \notin P_i$,[4] whence

$$P(a_i b_i p) = p^\varphi \quad \text{and} \quad P(a_i b_i q) = q^\varphi;$$

by des′ this implies $pq \| p^\varphi q^\varphi$. Consequently φ is a translation. Finally $a_1^\varphi = b_1$, because, by 2.4.4, $P(a_1 b_1 a_2 b_2)$ implies $P(a_2 b_2 a_1 b_1)$.

Proof of (ii) is analogous to that of (i); it is based on 2.3.11 and deals with the trapezoid relation. ■

As a corollary of 3.2.1 and 3.1.2 we obtain

Th 3.2.2. Des $\models\!\Rightarrow$ des.

This inference can be proved as well elementarily, without use of the notions of translation and homothety.

By 3.2.1 the set of sentences

$$\text{AX } 2.2 \cup \{\text{des}\}$$

is an axiom system for the translation planes, while the set

$$\text{AX } 2.2 \cup \{\text{Des}\}$$

is an axiom system for the homothety planes.[5]

We shall refer to translation planes and homothety planes as to *weakly Desarguean*[6] and *strongly Desarguean* (or briefly *Desarguean*) *affine planes*, respectively.[7] The class of all the translation planes will be denoted by the symbol des $\mathscr{A}$, the class of all the homothety planes—by the symbol Des $\mathscr{A}$.

3.3. PAPPUS'S LAW

Given a parallelity plane $\mathfrak{A} = (S, \|)$, let us consider the translation groups Tr_P and the homothety groups Ht_o of $\mathfrak{A}$. If we adjoin des to the

[4] In this point of the proof it becomes clear why it was convenient to deal with as many as three lines P_1, P_2, P_3 to define p^φ.

[5] The system AX 2.2 describes a parallelity plane $(S, \|)$. The atomic formulas for collinearity occuring in the axioms des and Des have to be treated as abbreviations of corresponding atomic formulas for parallelity (see remark at the end of § 2.2).

[6] The adjective *Desarguean* comes from the name *Desargues*, like the adjective *Euclidean* comes from the name *Euclid*.

[7] The term *Desarguean* for affine planes with transitive set of homotheties is in common use. As an analogue, we introduce the term *weakly Desarguean* for affine planes with transitive set of translations.

axioms for $\mathfrak{A}$ then, by 3.2.1 (i) and 3.1.1, each group Tr_P turns out to be transitive and commutative. If we adjoin Des, then by 3.2.1 (ii) each group Ht_o turns out to be transitive, but not necessarily commutative.[8] To obtain simultaneously transitivity and commutativity of the homothety groups, we need the axiom of Pappus which is stronger than Des:

Ax of Pappus $\quad L(p_1p_2p_3) \wedge L(q_1q_2q_3) \wedge p_1q_2 \| p_2q_1 \wedge p_2q_3 \| p_3q_2$
$$\Rightarrow p_1q_3 \| p_3q_1 .$$

This axiom can be reformulated in terms of lines as

Pappus's law $\quad p_1, p_2, p_3 \in P \wedge q_1, q_2, q_3 \in Q \wedge p_1q_2 \| p_2q_1$
$$\wedge p_2q_3 \| p_3q_2 \Rightarrow p_1q_3 \| p_3q_1 .$$

The following minor and major Pappus's laws, papp and Papp,[9] are particular cases of that law (see figs. 18 and 19):

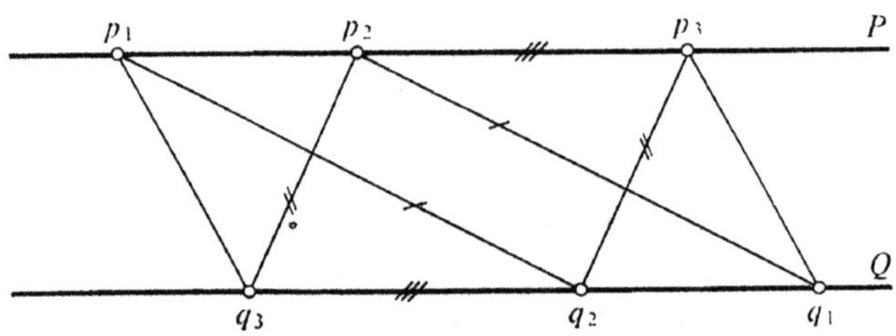

Fig. 18

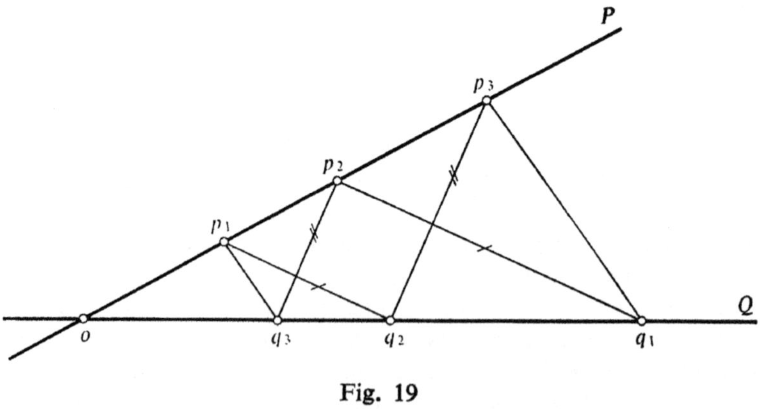

Fig. 19

[8] In § 4.0 we shall construct a Desarguean affine plane with non-commutative homothety groups.

[9] Adjoining inessential assumption $\neq (p_1p_2p_3)$, we can express papp and Papp

papp $P \| Q \wedge P \neq Q \wedge p_1, p_2, p_3 \in P \wedge q_1, q_2, q_3 \in Q$
$$\wedge p_1 q_2 \| p_2 q_1 \wedge p_2 q_3 \| p_3 q_2 \Rightarrow p_1 q_3 \| p_3 q_1 .$$

Papp $o \in P, Q \wedge P \neq Q \wedge o \neq p_1, p_2, p_3 \in P \wedge o \neq q_1, q_2, q_3 \in Q$
$$\wedge p_1 q_2 \| p_2 q_1 \wedge p_2 q_3 \| p_3 q_2 \Rightarrow p_1 q_3 \| p_3 q_1 .$$

Th 3.3.0. Pappus's law $\rightleftarrows$ papp $\wedge$ Papp.[10]

Proof. The inference $\Rightarrow$ is obvious. To prove $\Leftarrow$ it suffices to deduce Pappus's law from the axioms for $\mathfrak{A}$ in those two cases which are not included in papp and Papp:

(i) $P = Q$; then, by 2.2.15, the consequent of Pappus's law follows from its antecedent.

(ii) $P \cap Q = \{o\}$ and at least one of the points $p_1, p_2, p_3, q_1, q_2, q_3$ coincides with the point o. Since the Pappus's law is symmetric with respect to the indices 1 and 3 as well as with respect to the triples $(p_1 p_2 p_3)$ and $(q_1 q_2 q_3)$,[11] we may assume that either p_1 or p_2 coincides with o. By the assumptions of the Pappus's law

$$p_1 = o \Rightarrow o q_2 \| p_2 q_1 \Rightarrow q_2 = o \vee p_2 = o \Rightarrow p_2 q_3 \| p_3 o \vee o q_3 \| p_3 q_2$$
$$\Rightarrow q_3 = o \vee p_3 = o,$$
$$p_2 = o \Rightarrow p_1 q_2 \| o q_1 \wedge o q_3 \| p_3 q_2 \Rightarrow (p_1 = o \vee q_1 = o)$$
$$\wedge (q_3 = o \vee p_3 = o).$$

Thus, in both cases, at least one of the following conditions is satisfied:

$$p_1 = q_3 = o \quad \text{or} \quad p_3 = q_1 = o \quad \text{or} \quad p_1 = p_3 = o \quad \text{or}$$
$$q_1 = q_3 = o.$$

This yields $p_1 q_3 \| p_3 q_1$, which is the consequent of Pappus's law. ■

---

in terms of the parallelogram relation and the trapezoid relation, respectively:

papp″ $P(p_1 p_2 q_2 q_1) \wedge P(p_2 p_3 q_3 q_2) \wedge p_1 \neq p_3 \wedge L(p_1 p_2 p_3)$
$$\Rightarrow P(p_1 p_3 q_3 q_1).$$

Papp″ $T_0(p_1 p_2 q_2 q_1) \wedge T_0(p_2 p_3 q_3 q_2) \wedge p_1 \neq p_3 \Rightarrow T_0(p_1 p_3 q_3 q_1).$

(compare the laws des″ and Des″, § 3.2).

[10] The symbol " $\wedge$ " is used here to denote the operation assigning to every two formulas their conjunction.

[11] I.e., interchanging the numerals 1 and 3 or the letters p and q, we obtain a formula equivalent to Pappus's law on the basis of the axioms for affine planes.

Let us now reformulate Desargues's laws, to make them formally close to Pappus's law:[12]

$$\text{des}\quad P_1 \| P_2 \| P_3 \wedge P_1 \neq P_2 \neq P_3 \wedge a_1, b_1 \in P_1 \wedge a_2, b_2 \in P_2 \wedge a_3, b_3 \in P_3 \wedge a_1 a_2 \| b_1 b_2 \wedge a_2 a_3 \| b_2 b_3 \Rightarrow a_1 a_3 \| b_1 b_3.$$

$$\text{Des}\quad o \in P_1, P_2, P_3 \wedge P_1 \neq P_2 \neq P_3 \wedge o \neq a_1, b_1 \in P_1 \wedge o \neq a_2, b_2 \in P_2 \wedge o \neq a_3, b_3 \in P_3 \wedge a_1 a_2 \| b_1 b_2 \wedge a_2 a_3 \| b_2 b_3 \Rightarrow a_1 a_3 \| b_1 b_3.$$

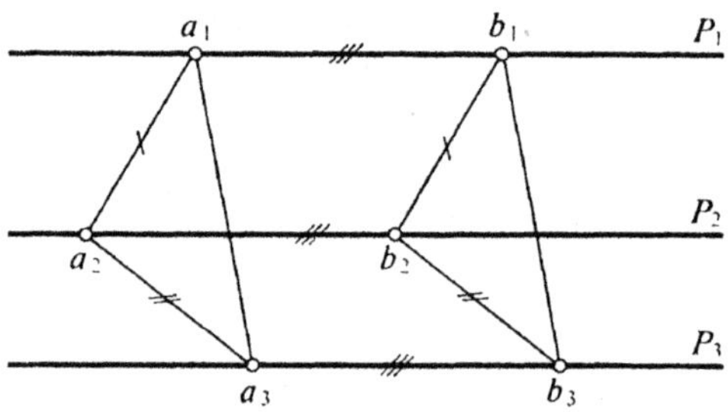

Fig. 20

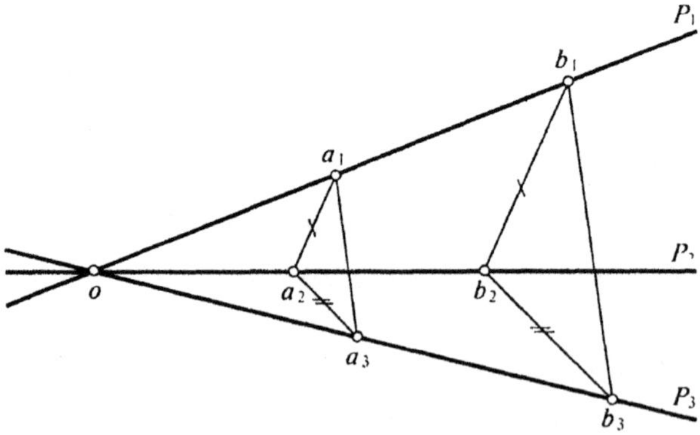

Fig. 21

[12] To make the changes in formulation clear, compare figures 15 and 16 with 20 and 21. In the last two, the degenerate case $P_1 = P_3$ is already neglected.

The laws papp and Papp, as well as des and Des, deal with some parallelities for six points. However, in the case of Pappus laws these points lie in threes on two lines, while in the case of Desargues's laws they lie in twos on three lines. We are now going to prove a theorem fundamental for affine planes.

Th 3.3.1. Pappus's law $\models\!\Rightarrow$ Des.

P r o o f. Let $\mathfrak{A}$ be a parallelity plane, and let us assume that Pappus's law is satisfied in $\mathfrak{A}$. The proof of Des will be indirect. Suppose that certain lines P_1, P_2, P_3, and certain points $a_1, a_2, a_3, b_1, b_2, b_3$ of the plane satisfy the conditions:

$$o \in P_1, P_2, P_3, \quad P_1 \neq P_2 \neq P_3,$$
$$o \neq a_1, b_1 \in P_1, \quad o \neq a_2, b_2 \in P_2, \quad o \neq a_3, \; b_3 \in P_3,$$
$$a_1 a_2 || b_1 b_2 \quad \text{and} \quad a_2 a_3 || b_2 b_3,$$

and

$$\sim (a_1 a_3 || b_1 b_3)$$

(see fig. 22). Then

$$P_1 \neq P_3,$$
$$\sim L(a_1 a_2 a_3) \quad \text{and} \quad \sim L(b_1 b_2 b_3),$$
$$\sim ([a_1 a_3] || P_2) \quad \text{or} \quad \sim ([b_1 b_3] || P_2).$$

Since Des is symmetric with respect to the triples $(a_1 a_2 a_3)$ and $(b_1 b_2 b_3)$, without any restriction of generality of the proof we may assume that

$$\sim ([a_1 a_3] || P_2).$$

We construct in turn auxiliary lines and points:

$$K = \mathbf{L}_{||[a_1 a_3]}(o),$$
$$\{p\} = K \cap [a_2 a_3],$$
$$L = \mathbf{L}_{||[a_1 a_2]}(p),$$
$$\{q\} = P_1 \cap L.$$

As easily seen

$$o \neq p \neq b_2 \quad \text{and} \quad q \neq a_3.$$

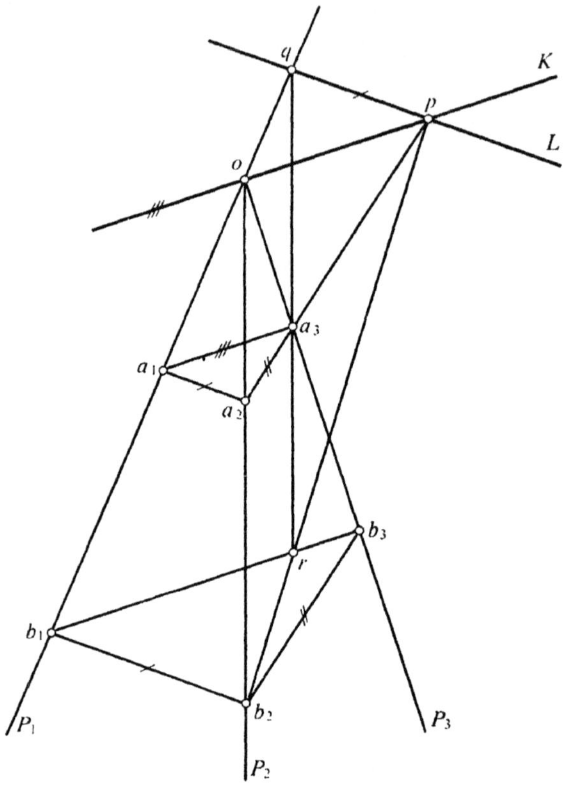

Fig. 22

We have also

$$o, a_1, q \in P_1 \wedge a_3, p, a_2 \in [a_2 a_3] \wedge op || a_1 a_3 \wedge a_1 a_2 || qp,$$

which, by Pappus's law, yields $oa_2 || qa_3$.[13] Hence

$$P_2 || [qa_3]$$

and consequently the lines $[pb_2]$ and $[qa_3]$ have a point of intersection r. It is easy to check that

$$o, q, b_1 \in P_1 \wedge r, b_2, p \in [pb_2] \wedge ob_2 || qr \wedge qp || b_1 b_2$$

and

$$o, a_3, b_3 \in P_3 \wedge r, b_2, p \in [pb_2] \wedge ob_2 || a_3 r \wedge a_3 p || b_3 b_2 .$$

[13] Since both cases $P_1 || [a_2 a_3]$ and $\sim P_1 || [a_2 a_3]$ are possible, we must use the whole Pappus's law.

Therefore, using twice Pappus's law, we get

$$op||b_1 r \quad \text{and} \quad op||b_3 r.$$

Hence $L(b_1 b_3 r)$ and $a_1 a_3||op||b_1 b_3$. As a result, Des is satisfied in $\mathfrak{A}$. ∎

Th 3.3.2.[14] Papp $\Rightarrow$ papp.

Proof. Let $\mathfrak{A}$ be a parallelity plane. Assume that Papp is satisfied in $\mathfrak{A}$. The proof of papp will be indirect. Suppose that certain lines P, Q and points $p_1, p_2, p_3, q_1, q_2, q_3$ of $\mathfrak{A}$ satisfy the following conditions:

$$P||Q, \ P \neq Q, \ p_1, p_2, p_3 \in P, \quad q_1, q_2, q_3 \in Q,$$
$$p_1 q_2||p_2 q_1 \quad \text{and} \quad p_2 q_3||p_3 q_2,$$

and

$$\sim (p_1 q_3||p_3 q_1).$$

It is easy to check that $\neq (p_1, p_2 p_3)$, $\neq (q_1 q_2 q_3)$ and $\sim p_1 q_3 ||p_3 q_2$. Let us construct two auxiliary lines and one point:

$$R = \mathbf{L}_{||[p_1 q_3]}(q_1),$$
$$\{p_3'\} = R \cap [p_3 q_2],$$
$$P' = [p_3' p_2].$$

Evidently

$$p_1 q_3||p_3' q_1 \quad \text{and} \quad p_3' \neq q_1.$$

Notice that

$$\sim (P'||Q) \quad \text{and} \quad \sim (P'||[p_1 q_2]).$$

Thus the line P' intersects the lines Q and $[p_1 q_2]$ in some points o and p_1' respectively. Then

$$o \in P', Q \wedge P' \neq Q \wedge o \neq p_1', p_2, p_3' \in P' \wedge o \neq q_1, q_2, q_3 \in Q$$
$$\wedge p_1' q_2||p_2 q_1 \wedge p_2 q_3||p_3' q_2,$$

which, by Papp, yields

$$p_1' q_3||p_3' q_1.$$

[14] This theorem is due to L. Włodarski (see Włodarski [79]). It was adjoined later together with the next two (M.M).

Consequently $p_1' q_3 || p_1 q_3$, whence $p_1' \in [p_1 q_2] \cap [p_1 q_3]$; thus $p_1' = p_1$ and $P' = P$, which implies $\sim (P||Q)$. As a result, papp is satisfied in $\mathfrak{A}$. ■

As an immediate consequence of 3.3.0 and 3.3.2, we get

Th 3.3.1. Papp $\rightleftarrows$ Pappus's law.

By 3.3.1 and 3.3.3 we get

Th 3.3.4. Papp $\Rightarrow$ Des.

We are now going to establish the relationships between the Pappus laws and some properties of the groups Tr_P and Ht_o.[15,16]

Th 3.3.5.

(i) *Each group* Tr_P *is transitive and commutative* $\Rightarrow$ *papp.*

(ii) *Each group* Ht_o *is transitive and commutative* $\Rightarrow$ Papp.

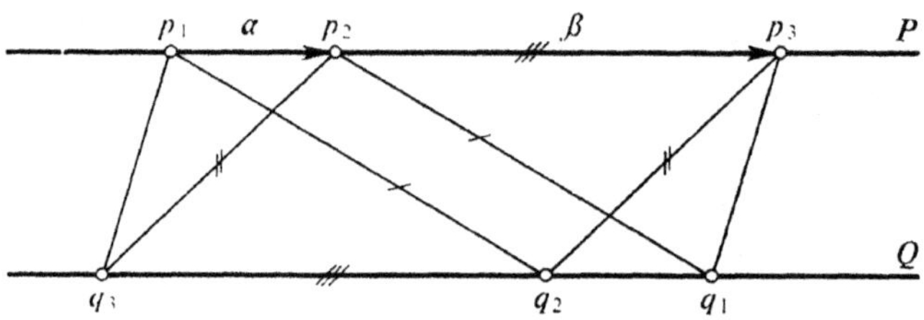

Fig. 23

Proof. (i). Assume all the groups Tr_P to be transitive and commutative. Let the lines P and Q and points $p_1, p_2, p_3, q_1, q_2, q_3$ satisfy the assumptions of papp (fig. 23):

$$P||Q, \quad P \neq Q, \quad p_1, p_2, p_3 \in P, \quad q_1, q_2, q_3 \in Q,$$
$$p_1 q_2 || p_2 q_1, \quad \text{and} \quad p_2 q_3 || p_3 q_2.$$

We have to show that $p_1 q_3 || p_3 q_1$. By the assumptions, there are $\alpha, \beta \in \mathsf{Tr}$

---

[15] To make these connections clearer, we shall formulate them concisely. We also omit the quotation-marks, which should be used for the sentences concerning the properties of Tr_P and Ht_o. Of course, the theorems 3.3.5, 3.3.6, and 3.3.8 might be expressed in the same style as 3.2.1.

[16] In 3.3.5(i), the condition *each group* Tr_P *is transitive and commutative* can be, of course, replaced by the condition *the group* Tr *is transitive* (see 3.1.1). Similarly, in 3.3.6(i), the condition *each group* Tr_P *is commutative* can be replaced by the condition *the group* Tr *is commutative* (see 2.5.19).

such that

$$p_1^{\alpha} = p_2 \quad \text{and} \quad p_2^{\beta} = p_3$$

and simultaneously

$$q_2^{\alpha} = q_1 \quad \text{and} \quad q_3^{\beta} = q_2 .$$

Thus

$$p_1^{\alpha+\beta} = p_3 \quad \text{and} \quad q_3^{\alpha+\beta} = q_3^{\beta+\alpha} = q_1 ,$$

and consequently $p_1 q_3 || p_3 q_1$.

The proof of (ii) is analogous to that of (i). ■

Th 3.3.6.

(i) papp ⊨⇒ *Each group* Tr_P *is commutative.*

(ii) Papp ⊨⇒ *Each group* Ht_o *is commutative.*

Proof. (i): Given a line P, let us take a line Q and two points p and q satisfying the conditions

$$(1) \qquad \begin{aligned} &P || Q \quad \text{and} \quad P \neq Q, \\ &p \in P \quad \text{and} \quad q \in Q \end{aligned}$$

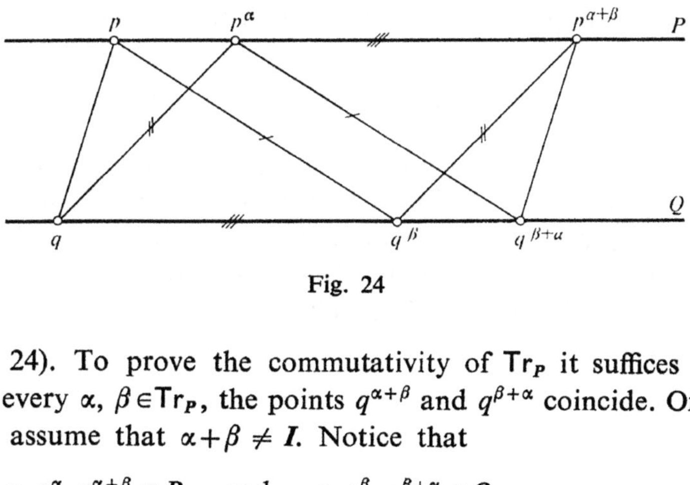

Fig. 24

(see fig. 24). To prove the commutativity of Tr_P it suffices to show that for every $\alpha, \beta \in \mathsf{Tr}_P$, the points $q^{\alpha+\beta}$ and $q^{\beta+\alpha}$ coincide. Of course, we may assume that $\alpha+\beta \neq \boldsymbol{I}$. Notice that

$$(2) \qquad p, p^{\alpha}, p^{\alpha+\beta} \in P \quad \text{and} \quad q, q^{\beta}, q^{\beta+\alpha} \in Q,$$

$$(3) \qquad pq^{\beta} || p^{\alpha} q^{\beta+\alpha} \quad \text{and} \quad p^{\alpha} q || p^{\alpha+\beta} q^{\beta}.$$

From (1), (2), and (3), by papp, it follows that $pq || p^{\alpha+\beta} q^{\beta+\alpha}$, which yields $\boldsymbol{P}(pp^{\alpha+\beta} qq^{\beta+\alpha})$. Thus $q^{\alpha+\beta} = \boldsymbol{P}(pp^{\alpha+\beta} q) = q^{\beta+\alpha}$.

The proof of (ii) is analogous to that of (i). ■

As corollaries, we derive the following two statements:

Th 3.3.7. Papp $\Rightarrow$ Des $\Rightarrow$ des $\Rightarrow$ papp.

Proof: by 3.3.4, 3.2.2, 3.2.1, 3.1.1, 3.3.5 (i). ■

As we shall show in the sequel, all these inferences are sharp, i.e. their converses are false.

Th 3.3.8. *Each group* Ht_o *is transitive and commutative* $\mathrel{|\!\rightleftarrows\!|}$ Papp.

Proof. Using 3.2.1 (ii), 3.3.5 (ii), 3.3.6 (ii), 3.3.7, we infer that

Each group Ht_o is transitive and commutative $\mathrel{|\!\rightleftarrows\!|}$ Des $\wedge$ Papp $\mathrel{|\!\rightleftarrows\!|}$ Papp. ■

If each group Ht_o of the parallelity plane $\mathfrak{A}$ is transitive and commutative, then $\mathfrak{A}$ is said to be a *Pappian affine plane* or a *commutative homothety plane.* Thus Pappian affine planes are those Desarguean planes which have commutative homothety groups. By theorem 3.3.8. the set of sentences

$$\text{AX } 2.2 \cup \{\text{Papp}\}$$

is an axiom system for Pappian affine planes. The class of all the commutative homothety planes will be denoted by the symbol Papp $\mathscr{A}$.

3.4. TRANSLATION PLANES

Let $\mathfrak{A} = (\boldsymbol{S}, \|)$ be a translation plane. By the definition, its *translation group is transitive* and thus, by 3.1.1, also *commutative.* By the transitivity of Tr, for any points $a, b \in \boldsymbol{S}$ there is a translation α which carries a onto b; by 2.5.12 (i), α is uniquely determined by these two points. Thus we have a function $\tau : \boldsymbol{S} \times \boldsymbol{S} \to \mathsf{Tr}$ with

Def 3.4.1. $a^{\tau(ab)} = b.$

Th 3.4.2.[17] $\tau(aa^{\alpha}) = \alpha.$
des

---

[17] All the theorems of §§ 2.2–3.3 concern an arbitrary parallelity plane. Now, we shall consider weakly Desarguean, strongly Desarguean, and Pappian planes. The terms "$\underset{\text{des}}{\text{Th}}$", "$\underset{\text{Des}}{\text{Th}}$", and "$\underset{\text{Papp}}{\text{Th}}$" will indicate what plane the theorem concerns.

Let us now fix an arbitrary point $o \in \boldsymbol{S}$. We introduce the operation $+_o$ (addition) on $\boldsymbol{S}$, dependent on the parameter o (fig. 25):[18]

Def 3.4.3. $p +_o q = o^{\tau(op)+\tau(oq)}$.

Definition 3.4.3 together with 3.4.1 yields

Th 3.4.4. $p +_o q = p^{\tau(oq)}$.
des

Consider the algebra $(\boldsymbol{S}, o, +_o)$. Let the function $\tau_o : \boldsymbol{S} \to \mathsf{Tr}$ be defined by

$$\tau_o(p) = \tau(op).$$

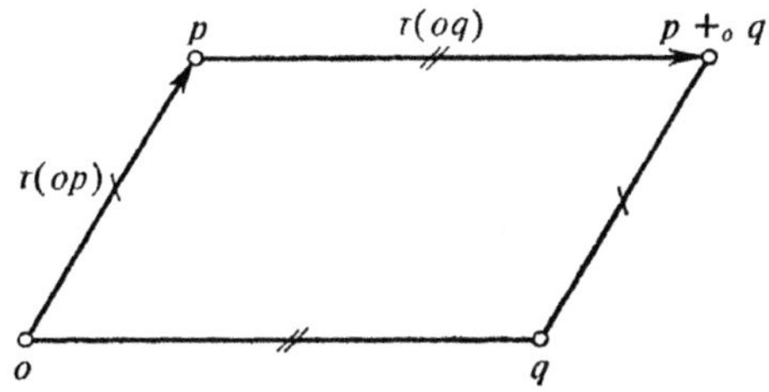

Fig. 25

By 3.4.2, the function τ_o is a bijection; by 3.4.3 and 3.4.2

$$\tau_o(p +_o q) = \tau(oo^{\tau(op)+\tau(oq)}) = \tau_o(p) + \tau_o(q)$$

and

$$\tau_o(o) = \tau(oo^I) = \boldsymbol{I};$$

thus τ_o maps $(\boldsymbol{S}, o, +_o)$ isomorphically onto the group $(\mathsf{Tr}, \boldsymbol{I}, +)$. Hence we have

Th 3.4.5. *The algebra* $(\boldsymbol{S}, o, +_o)$ *is a commutative group; it is iso-*
des

morphic to $(\mathsf{Tr}, \boldsymbol{I}, +)$.

Let us notice that the isomorphism τ_o carries a line $P \ni o$ onto the set Tr_P. Hence, the algebra $(P, o, +_o)$ is a subgroup of $(\boldsymbol{S}, o, +_o)$, isomorphic to the subgroup $(\mathsf{Tr}_P, \boldsymbol{I}, +)$ of $(\mathsf{Tr}, \boldsymbol{I}, +)$. Thus, we have

---

[18] If $\sim \boldsymbol{L}(opq)$, then obviously $p +_o q = \boldsymbol{P}(opq)$; however the operation $+_o$ is, contrary to $\boldsymbol{P}$, everywhere defined, and in particular the sum $p +_o q$ is defined for points p and q of arbitrary line P passing through o.

Th 3.4.6. *For every line P and every point $o \in P$, the algebra $(P, o, +_o)$*
des

is a commutative group; it is isomorphic to $(\mathrm{Tr}_P, I, +)$.

We shall refer to $(P, o, +_o)$ as the *additive group of the line P with o as zero.*

We shall now prove that, up to an isomorphism, the group $(P, o, +_o)$ is independent of P and o.

Th 3.4.7. *If*
des

(1) $K||L, \ o_1 \in K, \ o_2 \in L,$

then

$$(K, o_1, +_{o_1}) \simeq (L, o_2, +_{o_2}),$$

and the translation $\tau(o_1 o_2)$ is an isomorphism.

Proof.[19] Assume (1) (fig. 26) and let

$$\pi = \tau(o_1 o_2).$$

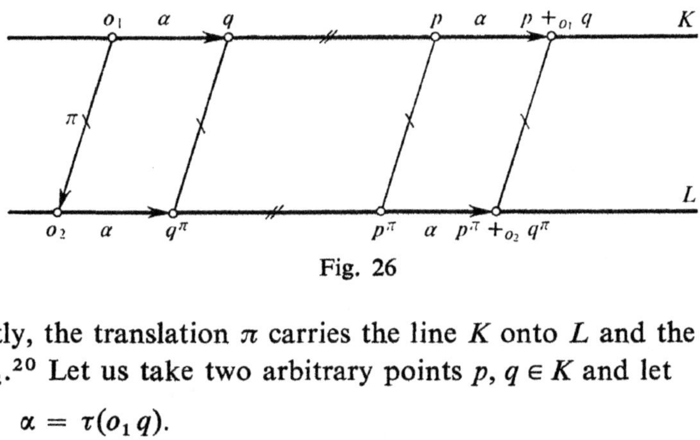

Fig. 26

Evidently, the translation π carries the line K onto L and the point o_1 onto o_2.[20] Let us take two arbitrary points $p, q \in K$ and let

$$\alpha = \tau(o_1 q).$$

Then

$$o_2^{\alpha} = o_1^{\pi\alpha} = o_1^{\alpha\pi} = q^{\pi},$$

i.e.

---

[19] It follows immediately from 2.5.8(i) and 3.4.6 that these two groups are isomorphic. However it is essential to find an isomorphism.

[20] It is clear that if $K \neq L$ then the translation $\tau(o_1 o_2)$ restricted to the line K is the parallel projection $\pi: K \underset{||[o_1 o_2]}{\rightarrow} L.$

$$\alpha = \tau(o_2 q^\pi);$$

since

$$p^{\alpha\pi} = p^{\pi\alpha},$$

by 3.4.4 we get

$$(p +_{o_1} q)^\pi = p^\pi +_{o_2} q^\pi,$$

which completes the proof. ■

Th 3.4.8. *If*
des

(2) $\quad K \cap L = \{o\},$

then

$$(K, o, +_o) \simeq (L, o, +_o),$$

and an arbitrary parallel projection $\pi: K \to L$ is an isomorphism.

P r o o f. Assume (2); fix a line P such that $\sim (P \| K)$ and $\sim (P \| L)$, and thus also a parallel projection $\pi: K \underset{\|P}{\to} L$. Of course

(3) $\quad o^\pi = o.$

Let us take any two points $p, q \in K$, and let

(4) $\quad \alpha = \tau(oq) \quad$ and $\quad \beta = \tau(oq^\pi).$

Then

(5) $\quad o^\alpha = q \quad$ and $\quad o^\beta = q^\pi,$

and, by 2.5.9 (ii),

$$\alpha \in \mathsf{Tr}_K \quad \text{and} \quad \beta \in \mathsf{Tr}_L.$$

We have to verify the condition

$$(p +_o q)^\pi = p^\pi +_o q^\pi,$$

which is, by 3.4.4, equivalent to

(*) $\quad p^{\alpha\pi} = p^{\pi\beta}.$

By (5), the condition (*) is satisfied, when $p = o$; it is also satisfied when $q = o$, because then $\alpha = \beta = I$. Let us assume $p, q \neq o$.

Consider first the particular case

(6) $\quad p^\alpha = o$

(fig. 27). From (5) and (6), by 2.5.20, it follows that

$$qq^\pi || po^{-\beta};$$

simultaneously

$$qq^\pi || pp^\pi$$

and $q \neq q^\pi$, whence $L(pp^\pi o^{-\beta})$. Since $o \neq p \in K$ and $p^\pi, o^{-\beta} \in L$, we get $p^\pi = o^{-\beta}$. Thus $p^{\pi\beta} = o$, which together with (6) and (3) gives (*).

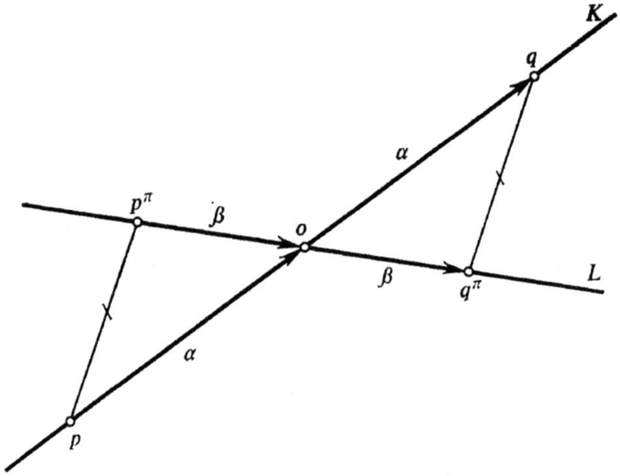

Fig. 27

We pass to the case

$$p^\alpha = o$$

(fig. 28). Evidently $p^\alpha \in K$ and

(7) $\quad pp^\pi || qq^\pi \quad$ and $\quad pp^\pi || p^\beta p^{\pi\beta};$

by 2.5.21, (5) implies

(8) $\quad p^\alpha p^\beta || qq^\pi.$

Since $p \neq p^\pi$ and $q \neq q^\pi$, from (7) and (8) it follow that

$$pp^\pi || p^\alpha p^\beta \quad \text{and} \quad L(p^\alpha p^\beta p^{\pi\beta}),$$

which together with $p^\alpha \neq p^\beta$ yields

$$pp^\pi || p^\alpha p^{\pi\beta}.$$

Simultaneously

$$pp^\pi || p^\alpha p^{\alpha\pi}$$

and $p \neq p^\pi$, thus $L(p^\alpha p^{\alpha\pi} p^{\pi\beta})$. Since $o \neq p^\alpha \in K$ and $p^{\alpha\pi}, p^{\pi\beta} \in L$, we get (∗). ∎

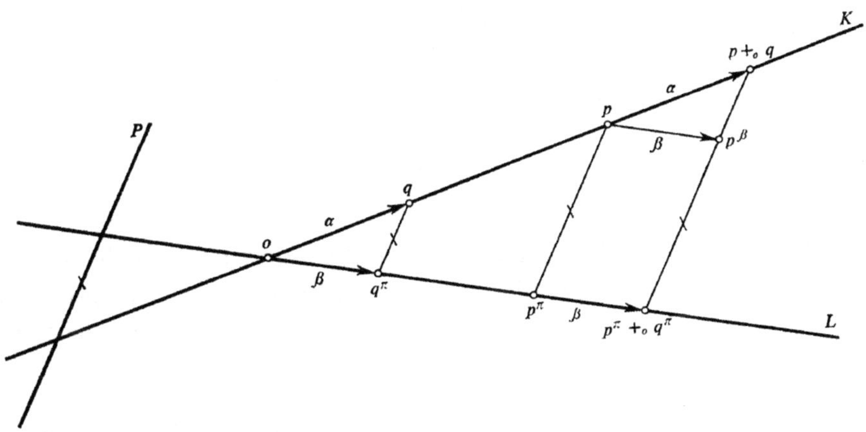

Fig. 28

We can now establish

Th 3.4.9. *For arbitrary lines K and L and arbitrary points* $o_1 \in K$
des
and $o_2 \in L$

$$(K, o_1, +_{o_1}) \simeq (L, o_2, +_{o_2}).$$

Proof. Let $o_1 \in K$ and $o_2 \in L$. By 3.4.7, we may assume $\sim (K||L)$. Let $K \cap L = \{o\}$. Then, by 3.4.7 and 3.4.8

$$(K, o_1, +_{o_1}) \simeq (K, o, +_o) \simeq (L, o, +_o) \simeq (L, o_2, +_{o_2}).$$ ∎

By theorems 3.4.6 and 3.4.9, *all the groups* Tr_P *of a translation plane* $\mathfrak{A}$ *are isomorphic.*

As a direct consequence of 3.4.8 we get

Th 3.4.10. *For an arbitrary line P and arbitrary points $o \in P$ and*
des
$a, b \in P - \{o\}$, there is an automorphism π of the group $(P, o, +_o)$, carrying a onto b.

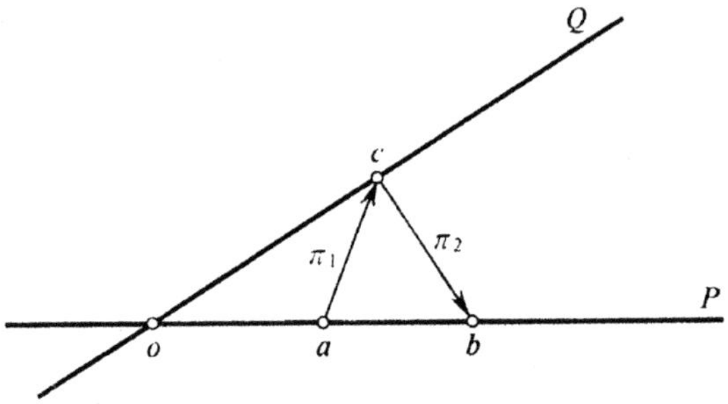

Fig. 29

P r o o f. Let us choose a line Q and a point $c \in Q$ such that $P \cap Q = \{o\}$ and $c \in Q - \{o\}$ (fig. 29). Consider two parallel projections $\pi_1 : P \underset{\|[ac]}{\rightarrow} Q$ and $\pi_2 : Q \underset{\|[cb]}{\rightarrow} P$. By 3.4.8, their composition $\pi = \pi_1 \pi_2$ is an automorphism of the group $(P, o, +_o)$, with $a^\pi = b$. ∎

By 3.4.10, the *group $(P, o, +_o)$ is homogeneous,* and thus *all its non-zero elements have the same order.* This holds also for the translation subgroup Tr_P which is isomorphic to $(P, o, +_o)$, and thus it also holds for the whole group Tr, since all its subgroups Tr_P are isomorphic. So, we have

Th 3.4.11. *All different from the identity elements of* Tr have *the*
des
same order.

3.5. HOMOTHETY PLANES

Let $\mathfrak{A} = (S, \|)$ be a homothety plane. By the definition, all its homothety groups Ht_o are transitive. Since $\mathfrak{A}$ is simultaneously a translation plane, all the translation groups Tr_P are transitive as well, and for each line P an arbitrary point $o \in P$ generates the commutative additive group $(P, o, +_o)$.

Let us fix a line P of $\mathfrak{A}$ and a point $o \in P$. By the transitivity of Ht_o, for every points $a, b \in P - \{o\}$ there is a homothety $\varkappa \in \mathrm{Ht}_o$ carrying a onto b. By 2.5.12 (ii), $\varkappa$ is determined uniquely by a and b. So, we have the function $\iota_o : (P - \{o\}) \times (P - \{o\}) \to \mathrm{Ht}_o$:

Def 3.5.1. $o \in P \wedge o \neq a, b \in P \Rightarrow o^{\iota_o(ab)} = o \wedge a^{\iota_o(ab)} = b.$

Th 3.5.2.[21] $o \in P \wedge o \neq a \in P \wedge o^{\varkappa} = 0 \Rightarrow \iota_o(aa^{\varkappa}) = \varkappa.$
Des

Given P and o, let us now fix also a point $a \in P - \{o\}$. We define an operation $\cdot^a_o$ (multiplication) on the set $P - \{o\}$, dependent on two parameters o and a (fig. 30[22]):

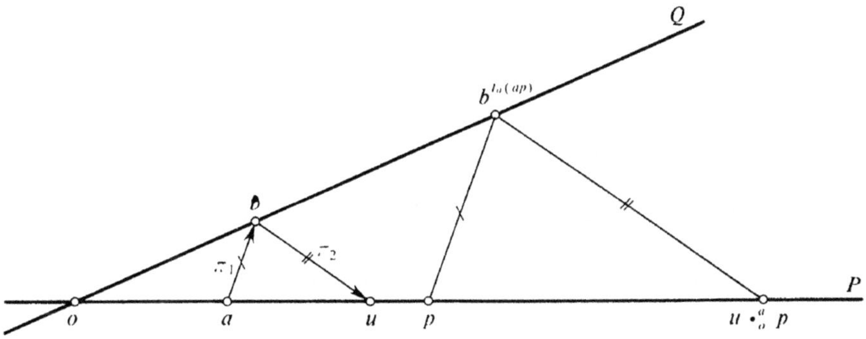

Fig. 30

Def 3.5.3. $o \in P \wedge o \neq a, p, q \in P \Rightarrow p \cdot^a_o q = a^{\iota_o(ap) \cdot \iota_o(aq)}.$

By 3.5.3 and 3.5.1 we have

Th 3.5.4. $o \in P \wedge o \neq a, p, q \in P \Rightarrow p \cdot^a_o q = p^{\iota_o(aq)}.$
Des

Consider the algebra $(P - \{o\}, a, \cdot^a_o)$ .Let $\iota^a_o : P - \{o\} \to \mathrm{Ht}_o$ be defined by

$$\iota^a_o(p) = \iota_o(ap).$$

By 3.5.2, the function ι^a_o is a bijection; by 3.5.3 and 3.5.2 this bijection maps $(P - \{o\}, a, \cdot^a_o)$ isomorphically onto the group $(\mathrm{Ht}_o, I, \cdot)$. Thus, we have

[21] See footnote 17.

[22] To construct the product $p \cdot^a_o q$, we choose an arbitrary line $Q \ni o$, different from P, and an arbitrary point $b \in Q - \{o\}$. Of course the point $p \cdot^a_o q$ is independent of the parameters Q and b.

Th 3.5.5. *For every line P and every two points $o \in P$ and $a \in P-\{o\}$*
Des
the algebra $(P-\{o\}, a, \cdot\,^{a}_{o})$ *is a group; it is isomorphic to* $(\mathrm{Ht}_o, I, \cdot)$.

We extend the operation $\cdot\,^{a}_{o}$ over the whole P by

Def 3.5.6. $o \in P \wedge o \neq a, p \in P \Rightarrow p \cdot\,^{a}_{o}\, o = o \cdot\,^{a}_{o}\, p = o.$

With this extended multiplication, the algebra $(P, o, a, \cdot\,^{a}_{o})$ is a multiplicative group with zero; it will be referred to as the *multiplicative group of the line P with o and a as zero and unit respectively.*

Let us pass to the problem of isomorphisms:

Th 3.5.7. *If*
Des

(1) $K \| L, \quad K \neq L, \quad o_1 \in K$ and $o_1 \neq a \in K,$
$o_2 \in L$ *and* $o_2 \neq b \in L, \quad o_1 o_2 \| ab,$

then

$$(K, o_1, a, \cdot\,^{a}_{o_1}) \simeq (L, o_2, b, \cdot\,^{b}_{o_2}),$$

and the translation $\tau(o_1 o_2)$ is an isomorphism.

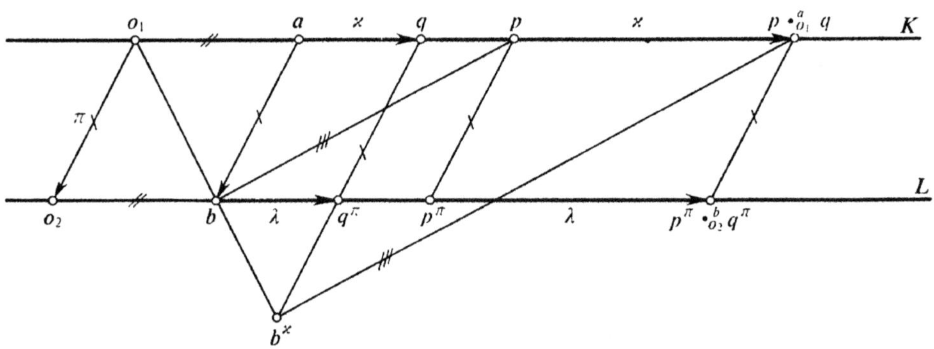

Fig. 31

Proof. Assume (1) (fig. 31) and let

$$\pi = \pi(o_1 o_2).$$

Evidently the translation[23] π carries the line K onto L, and

(2) $o_1^\pi = o_2$ and $a^\pi = b.$

---

[23] Of course, the translation $\tau(o_1 o_2)$ restricted to K coincides with the parallel projection $\pi: K \underset{\|[o_1 o_2]}{\rightarrow} L.$

Take now an arbitrary point $q \in K - \{o_1\}$ and let

$$\varkappa = \iota_{o_1}(aq) \quad \text{and} \quad \lambda = \iota_{o_2}(bq^\pi).$$

Then $\varkappa \in \mathrm{Ht}_{o_1}$, $\lambda \in \mathrm{Ht}_{o_2}$, and

$$a^\varkappa = q \quad \text{and} \quad b^\lambda = q^\pi.$$

Moreover

$$\pi^{-1}\varkappa\pi = \lambda,$$

since, by 2.5.4 (i), $\pi^{-1}\varkappa\pi \in \mathrm{Ht}_{o_2}$, and

$$b^{\pi^{-1}\varkappa\pi} = a^{\varkappa\pi} = q^\pi = b^\lambda.$$

Thus $\varkappa\pi = \pi\lambda$; in particular for any $p \in K - \{o_1\}$

$$p^{\varkappa\pi} = p^\pi ,$$

which, by 3.5.4, may be rewritten as

(3) $$(p \cdot {}^a_{o_1} q)^\pi = p^\pi \cdot {}^b_{o_2} q^\pi.$$

The formula (3) is valid for $p, q \in K - \{o_1\}$; by 3.5.6 it remains valid for $p = o_1$ and for $q = o_1$. By (2) and (3) the translation π maps isomorphically the multiplicative group of K onto the multiplicative group of L. ∎

Th 3.5.8. *If*
Des

(4) $$K \cap L = \{o\}, \quad o \neq a \in K, \quad o \neq b \in L,$$

then

$$(K, o, a, \cdot {}^a_o) \simeq (L, o, b, \cdot {}^b_o),$$

and the projection $\pi\colon K \underset{||[ab]}{\rightarrow} L$ *is an isomorphism.*

Proof. Assume (4) (fig. 32). Obviously

(5) $$o^\pi = o,$$

and

(6) $$a^\pi = b.$$

Take an arbitrary point $p \in K - \{o\}$ and a homothety $\varkappa \in \mathrm{Ht}_o$. Clearly $p^\varkappa \in K$ and

$$pp^\pi || p^\varkappa p^{\varkappa\pi} \quad \text{and} \quad pp^\pi || p^\varkappa p^{\pi\varkappa},$$

which, together with $p \neq p^\pi$, yields

$$L(p^\varkappa p^{\varkappa\pi} p^{\pi\varkappa});$$

since $o \neq p^\varkappa \in K$ and $p^{\varkappa\pi}, p^{\pi\varkappa} \in L$, it follows that

(*) $\quad p^{\varkappa\pi} = p^{\pi\varkappa}.$

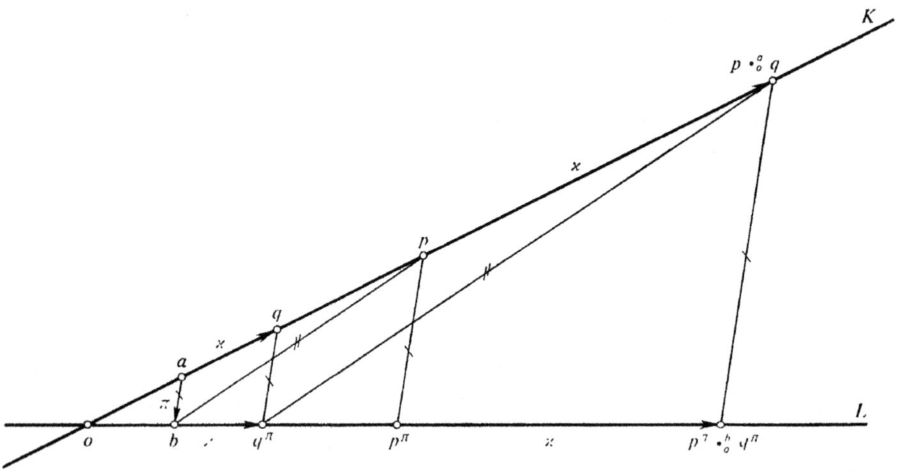

Fig. 32

In particular, by (6), for $p = a$ we get

(7) $\quad a^{\varkappa\pi} = b^\varkappa.$

Take now an arbitrary point $q \in K - \{o\}$ and let

(8) $\quad \varkappa = \iota_o(aq).$

Then

$$a^\varkappa = q,$$

which, together with (7), yields

$$b^\varkappa = q^\pi,$$

i.e.

(9) $\quad \varkappa = \iota_o(bq^\pi).$

By 3.4.5, using (8) and (9), we may rewrite (*) as

(10) $\quad (p \cdot^a_o q)^\pi = p^\pi \cdot^b_o q^\pi.$

The formula (10) is valid for $p, q \in K - \{o\}$; by 3.5.6 and (5) it remains valid for $p = o$ and for $q = o$. By (5), (6), and (10) the projection π maps isomorphically the multiplicative group of K onto that of L. ∎

Let us now expand the multiplicative group $(P, o, a, \cdot{}^{a}_{o})$ by the addition $+_o$. We have

Th 3.5.9. *For every line P and every two points $o \in P$ and $a \in P - \{o\}$,*
Des
the algebra $(P, o, a, +_o, \cdot{}^{a}_{o})$ is a field.

Proof. By 3.4.6, the subalgebra $(P, o, +_o)$ is a commutative group; by 3.5.5 and 3.5.6 the subalgebra $(P, o, a, \cdot{}^{a}_{o})$ is a multiplicative group with the point o as zero. Thus, it remains to verify the distributivity laws:

(I) $$(p +_o q) \cdot{}^{a}_{o} u = p \cdot{}^{a}_{o} u +_o q \cdot{}^{a}_{o} u,$$

(II) $$u \cdot{}^{a}_{o} (p +_o q) = u \cdot{}^{a}_{o} p +_o u \cdot{}^{a}_{o} q,$$

i.e. it remains to show that both right- and left-sided multiplications by a point u are automorphisms of the additive group $(P, o, +_o)$.

The right-sided multiplication by u coincides (on the line P) with the homothety $\varkappa = \iota_o(au)$, since by 3.5.4 for arbitrary $p \in P$

$$p \cdot{}^{a}_{o} u = p^{\varkappa}.$$

On the other hand, $\varkappa$ is an automorphism of the plane $\mathfrak{A}$, preserving P and o, therefore it is an automorphism of the group $(P, o, +_o)$ as well.[24]

[24] A different proof of the law (I) is based on 2.5.4(ii): Let

$$\alpha = \tau(oq), \quad \varkappa = \iota_o(au), \quad \beta = \tau(oq^{\varkappa}).$$

Then

$$o^{\alpha} = q, \quad o^{\varkappa} = o, \quad o^{\beta} = q^{\varkappa}.$$

Moreover

$$\varkappa^{-1}\alpha\varkappa = \beta,$$

because $\varkappa^{-1}\alpha\varkappa \in \mathsf{Tr}$ by 2.5.4(ii), and

$$o^{\varkappa^{-1}\alpha\varkappa} = o^{\alpha\varkappa} = q^{\varkappa} = o^{\beta}.$$

Thus $\alpha\varkappa = \varkappa\beta$ and, in particular, for any $p \in P$

$$p^{\alpha\varkappa} = p^{\varkappa\beta},$$

which, by 3.4.4 and 3.5.4, is equivalent to (I).

This proof may be found in the literature (compare e.g. Lingenberg–Baur [60], p. 125); evidently, our proof is much simpler.

Let us pass to the left-sided multiplication by u. Take a line Q and a point b satisfying the conditions:

$$P \cap Q = \{o\} \quad \text{and} \quad o \neq b \in Q$$

(fig. 33), and consider the parallel projections

$$\pi_1 : P \underset{||[ab]}{\rightarrow} Q \quad \text{and} \quad \pi_2 : Q \underset{||[bu]}{\rightarrow} P.$$

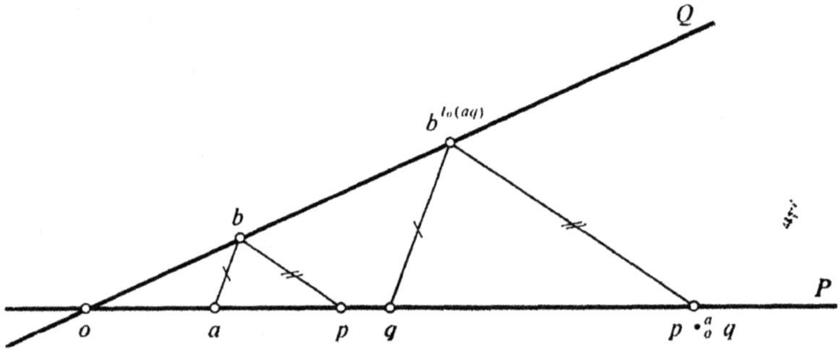

Fig. 33

Clearly

$$u \cdot^{a}_{o} p = p^{\pi_1 \pi_2},$$

and, by 3.4.8, the composition $\pi_1 \pi_2$ of the projections π_1 and π_2 is an automorphism of the group $(P, o, +_o)$. ■

The field $(P, o, a, +_o, \cdot^{a}_{o})$ will be referred to as the *field of the line P, with o as zero and a as unit*. We are going to show that its isomorphism type is independent of the parameters o and a:

Th 3.5.10. *For arbitrary lines K and L and arbitrary points* $o_1 \in K$,
Des

$a \in K - \{o_1\}$, $o_2 \in L$, $b \in L - \{o_2\}$

$$(*) \qquad (K, o_1, a, +_{o_1}, \cdot^{a}_{o_1}) \simeq (L, o_2, b, +_{o_2}, \cdot^{b}_{o_2}).$$

P r o o f. We consider three possibilities:

(i) $K \cap L = \{o\}$. If $o_1 = o_2 = o$, then by 3.4.8 and 3.5.8 the projection $\pi : K \underset{||[ab]}{\rightarrow} L$ maps isomorphically the field of K onto that of L. Assume in turn, that at least one of the points o_1 and o_2 is different from o; without any restriction of generality of the proof we may assume that

$o_2 \neq o$ (fig. 34). Taking the auxiliary line $M = \mathbf{L}_{\|K}(o_2)$ and auxiliary point $c = \boldsymbol{P}(o_1 a o_2) \in M$, we get

$$(11) \qquad (K, o_1, a, +_{o_1}, \cdot\,^{a}_{o_1}) \simeq (M, o_2, c, +_{o_2}, \cdot\,^{c}_{o_2}) \simeq (L, o_2, b, +_{o_2}, \cdot\,^{b}_{o_2});$$

indeed, by 3.4.7 and 3.5.7, the translation $\tau(o_1 o_2)$ maps isomorphically the field of K onto that of M, and by 3.4.8 and 3.5.8, the projection $\pi': M \underset{\|[cb]}{\to} L$ maps isomorphically the field of M onto that of L.

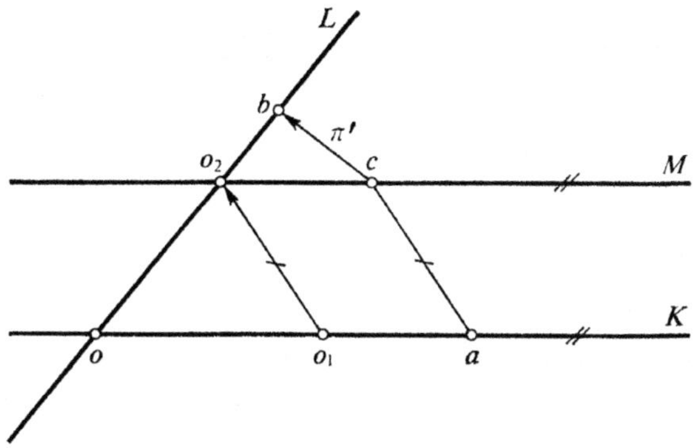

Fig. 34

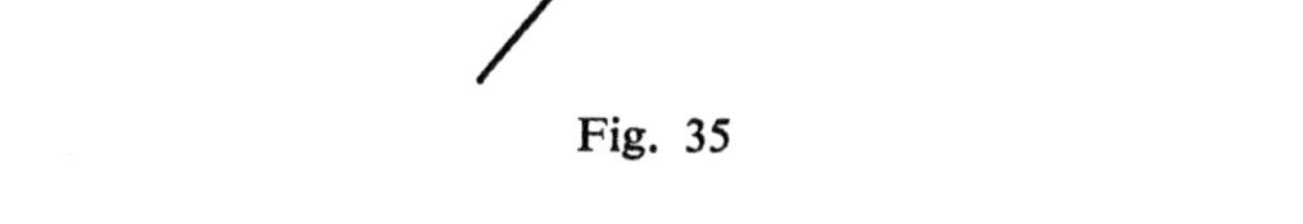

Fig. 35

(ii) $K = L$ (fig. 35). Take a line $M \ni o_2$ different from K and a point $c \in M - \{o_2\}$. By 3.5.10 (i)[25] we again get (11).

(iii) $K \| L$ and $K \neq L$ (fig. 36). Setting $M = L$ and choosing an auxiliary point $c = \boldsymbol{P}(o_1 a o_2) \in M$, we also get (11); indeed, by 3.4.7 and 3.5.7 the translation $\tau(o_1 o_2)$ maps isomorphically the field of K

[25] By 3.5.10(i) we understand the statement obtained from 3.5.10 by adjoining (i) to the assumptions.

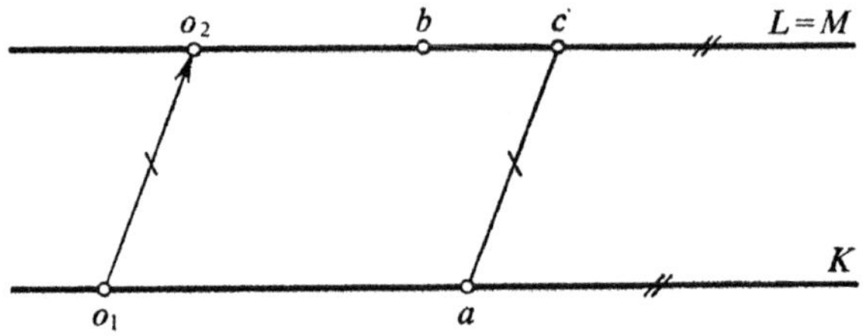

Fig. 36

onto that of M, and by 3.5.10 (ii) the field of M is isomorphic to that of L. ■

From (∗) we deduce in particular[26]

(∗∗) $(K-\{o_1\}, a, \cdot^{a}_{o_1}) \simeq (L-\{o_2\}, b, \cdot^{b}_{o_2})$.

Hence, by 3.5.5, *on the homothety plane* $\mathfrak{A}$ *all the homothety groups* Ht_o *are isomorphic.*

As a consequence of 3.5.9, 3.5.5, and 3.3.8 we get

Th 3.5.11. *For arbitrary line P and arbitrary points $o \in P$, $a \in P-\{o\}$*
Papp

the algebra $(P, o, a, +_o, \cdot^{a}_{o})$ is a commutative field.

[26] Of course, to prove that those two groups are isomorphic, it is not necessary to deal with fields. However, in both cases the same construction is needed for the proof.

CHAPTER 4

COORDINATIZATION AND ALGEBRAIC REPRESENTATION OF AFFINE PLANES

4.0. ALGEBRAIC PARALLELITY PLANES

In §§ 2.2, 3.2, and 3.3 we described axiomatically certain classes of parallelity planes:

$$\mathscr{A} \supseteq \text{des}\,\mathscr{A} \supseteq \text{Des}\,\mathscr{A} \supseteq \text{Papp}\,\mathscr{A}.$$

We shall now use some algebraic structures to construct examples of parallelity planes. The corresponding classes of algebras are:

$$\mathbf{TF} \underset{\text{Def}}{\supseteq} \mathbf{QF} \supseteq \mathbf{F} \supseteq \text{com}\,\mathbf{F}$$

(see § 0).

We start with the class of ternary fields, **TF** .To every ternary field $\mathfrak{T} = (F, 0, 1, \mathbf{T})$ we assign the structure

$$\underline{\mathfrak{A}}(\mathfrak{T}) = (F^2, ||_{\mathfrak{T}}),$$

setting for arbitrary points

$$a = (x_a, y_a), b = (x_b, y_b), c = (x_c, y_c), d = (x_d, y_d) \in F^2$$

Def 4.0.0. $ab||_{\mathfrak{T}}cd \Leftrightarrow x_a = x_b \wedge x_c = x_d$
$\vee \exists uvv' (\mathbf{T}(ux_av) = y_a \wedge \mathbf{T}(ux_bv) = y_b$
$\wedge \mathbf{T}(ux_cv') = y_c \wedge \mathbf{T}(ux_dv') = y_d).$[0]

Theorem 4.0.1. $\mathfrak{T} \in \mathbf{TF} \Rightarrow \underline{\mathfrak{A}}(\mathfrak{T}) \in \mathscr{A}.$

P r o o f. Let $\mathfrak{T} \in \mathbf{TF}$. Using in succesion the axioms
0.4.7; 0.4.7 and 0.4.5; 0.4.9, 0.4.6, 0.4.0; 0.4.7 and 0.4.5; 0.4.7 and 0.4.8 for ternary fields, we easily show that the axioms
2.2.0; 2.2.1; 2.2.2, 2.2.3, 2.2.4; 2.2.5; 2.2.6
for parallelity planes are satisfied in $\underline{\mathfrak{A}}(\mathfrak{T})$. ∎

We pass from the class of ternary fields, **TF**, to the class of strong left quasi-fields, **QF** (see 0.5.1). Let

[0] We admit the convention that $\wedge$ binds stronger than $\vee$.

$\mathfrak{F} = (F, 0, 1, +, \cdot) \in \mathbf{QF}$. Using 0.5.1, we may now reformulate the definition 4.0.0 as follows:

$$\begin{aligned} ab\|_{\mathfrak{F}}cd \Leftrightarrow x_a = x_b \wedge x_c = x_d \\ \vee \exists uvv' \, (u \cdot x_a + v = y_a \wedge u \cdot x_b + v = y_b \\ \wedge u \cdot x_c + v' = y_c \wedge u \cdot x_d + v' = y_d). \end{aligned}$$

By simple calculation in the left quasi-field $\mathfrak{F}$, we get

Theorem 4.0.2. *If* $\mathfrak{F} \in \mathbf{QF}$, *then*

$$\begin{aligned} ab\|_{\mathfrak{F}}cd \Leftrightarrow x_a = x_b \wedge x_c = x_d \vee \exists u \, (u \cdot (x_a - x_b) = y_a - y_b \\ \wedge u \cdot (x_c - x_d) = y_c - y_d). \end{aligned}$$

By 4.0.1, the structure

$$\underline{\mathfrak{A}}(\mathfrak{F}) = (F^2, \|_{\mathfrak{F}})$$

is a parallelity plane. We shall now examine the set of translations, $\mathrm{Tr}_{\mathfrak{F}}$, of $\underline{\mathfrak{A}}(\mathfrak{F})$. Let $+$ be the addition in the Cartesian square of the additive group $(F, 0, +)$ of $\mathfrak{F}$. Let us consider the transformation $\alpha: F^2 \to F^2$ defined for fixed $a \in F^2$ by the formula

$$p^\alpha = p + a.$$

Evidently α is a bijection, equal to the identity if $a = (0, 0)$, and without fixed points otherwise. Since

$$p - q = p^\alpha - q^\alpha,$$

by 4.0.2 it follows that

$$pq\|_{\mathfrak{F}}p^\alpha q^\alpha.$$

Thus α is a translation. The set of translations

$$\{\alpha: p^\alpha = p + a \text{ for some } a \in F^2\}$$

is transitive:

$$p^{\tau(ab)} = p + (b - a),$$

and thus it contains all the translations. Hence, we have

Theorem 4.0.3. *If* $\mathfrak{F} \in \mathbf{QF}$, *then*

(i) $\mathrm{Tr}_F = \{\alpha: p^\alpha = p + a$ *for some* $a \in F^2\}$.

(ii) *The set* Tr_F *is transitive.*

Let us now restrict our algebraic basis to the class of arbitrary fields, **F**. By theorems 4.0.2 and 0.6.4, the relation $\|_{\mathfrak{F}}$ can be described in terms of the right Cartesian linear space $\underline{\mathfrak{B}}^2(\mathfrak{F})$ over $\mathfrak{F}$:

Theorem 4.0.4. *If* $\mathfrak{F} \in \mathbf{F}$, *then*

$$ab \|_{\mathfrak{F}} cd \Leftrightarrow \exists uv\,\big((u, v) \neq (0, 0) \wedge (a-b)\cdot u = (c-d)\cdot v\big).$$

Let $o \in F^2$. We shall now examine the set of homotheties, $(\mathsf{Ht}_{\mathfrak{F}})_o$, of $\underline{\mathfrak{A}}(\mathfrak{F})$. Let us consider the transformation $\varkappa: F^2 \to F^2$ defined for a fixed $u \in F - \{o\}$ in terms of $\underline{\mathfrak{B}}^2(\mathfrak{F})$ by the formula

$$p^{\varkappa} - o = (p-o)\cdot u \qquad (u \neq 0).$$

Evidently $\varkappa$ is a bijection, equal to the identity if $u = 1$, and with the unique fixed point o otherwise. Since

$$(p-q)\cdot u = p^{\varkappa} - q^{\varkappa},$$

by 4.0.4 it follows that

$$pq \|_{\mathfrak{F}} p^{\varkappa} q^{\varkappa}.$$

Thus $\varkappa$ is a homothety with center o. The set of homotheties

$$\{\varkappa: p^{\varkappa} - o = (p-o)\cdot u \text{ for some } u \in F - \{o\}\}$$

is transitive. Indeed, if the points $a, b \in F^2$ satisfy the conditions

$$o \neq a, b \quad \text{and} \quad L_{\mathfrak{F}}(oab),$$

then, by 4.0.4, for some $u \neq 0$

$$b - o = (a-o)\cdot u$$

and

$$p^{\iota_o(ab)} - o = (p-o)\cdot u.$$

Thus we have

Theorem 4.0.5. *If* $\mathfrak{F} \in \mathbf{F}$, *then*

(i) $(\mathsf{Ht}_{\mathfrak{F}})_o = \{\varkappa: p^{\varkappa} - o = (p-o)\cdot u \textit{ for some } u \in F - \{o\}\}$.

(ii) *The set* $(\mathsf{Ht}_{\mathfrak{F}})_o$ *is transitive.*

On the basis of 4.0.3 (ii) and 4.0.5, we shall easily establish the connections between the classes of algebras **QF**, **F**, com **F** and the classes of parallelity planes des $\mathscr{A}$, Des $\mathscr{A}$, Papp $\mathscr{A}$:

Theorem 4.0.6. (i) $\mathfrak{F} \in \mathbf{QF} \Rightarrow \underline{\mathfrak{A}}(\mathfrak{F}) \in \text{des}\,\mathscr{A}$.[1]

(ii) $\mathfrak{F} \in \mathbf{F} \Rightarrow \underline{\mathfrak{A}}(\mathfrak{F}) \in \text{Des}\,\mathscr{A}$.

(iii) $\mathfrak{F} \in \text{com}\,\mathbf{F} \Rightarrow \underline{\mathfrak{A}}(\mathfrak{F}) \in \text{Papp}\,\mathscr{A}$.

---

[1] The author's conjecture was that if the axiom 0.2.2 str for quasi-fields is omitted then one gets pseudo-Euclidean affine planes, which can be expanded to pseudo-Euclidean planes. It would be one more advantage of our axiom system.

Proof. (i) follows immediately from 4.0.1 and 4.0.3 (ii), by virtue of 3.2.1 (i).

(ii) follows immediately from 4.0.1 and 4.0.5 (ii), by virtue of 3.2.1 (ii).

(iii) It suffices to show that the group $(\mathrm{Ht}_{\mathfrak{F}})_o$ is commutative (comp. 3.3.8). Indeed, if $\varkappa, \lambda \in (\mathrm{Ht}_{\mathfrak{F}})_o$, then by 4.0.5 (i) for some $u, v \in F$

$$p^{\varkappa} - o = (p-o) \cdot u \quad \text{and} \quad p^{\lambda} - o = (p-o) \cdot v,$$

which implies

$$p^{\varkappa\lambda} - o = (p-o) \cdot (u \cdot v) \quad \text{and} \quad p^{\lambda\varkappa} - o = (p-o) \cdot (v \cdot u).$$

From $u \cdot v = v \cdot u$ it follows that $p^{\varkappa\lambda} = p^{\lambda\varkappa}$ for every $p \in F^2$. ∎

4.1. COORDINATIZATION AND ITS PARAMETERS

In §§ 4.1-4.5 we shall show by means of coordinatization of a plane, that the class of all the parallelity planes coincides up to isomorphism with the class of the parallelity planes constructed above. Let $\mathfrak{A} = (S, \|)$ be an arbitrary parallelity plane. The coordinatization of $\mathfrak{A}$ is understood as the complex of the following three tasks: construction of a Cartesian coordinate system; construction of the scalar algebra on a coordinate

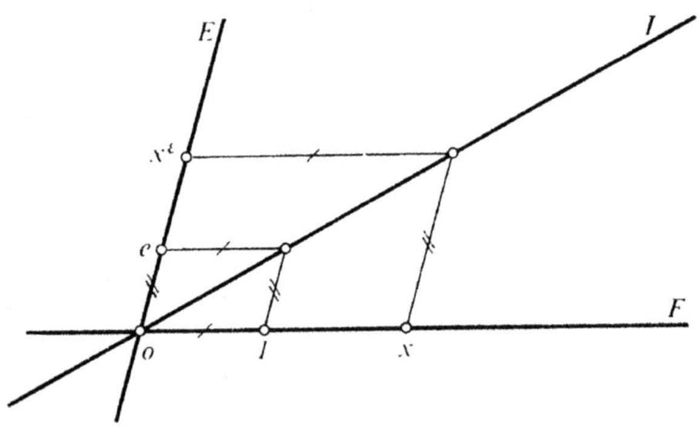

Fig. 37

axis of this system; algebraic description of the parallelity relation, i.e. expression of the formula $ab \| cd$ in terms of the coordinates of a, b, c, d, and the operations of the scalar algebra. Since for arbitrary parallelity plane we do not have the transitive groups of translations and homotheties, the construction of the coordinate system and of the scalar algebra must be based on the parallelogram and trapezoid operations, $\boldsymbol{P}$ and $\boldsymbol{T}$.

Let us fix three arbitrary non-collinear points o, 1, $e \in \boldsymbol{S}$:

Ass 4.1.0. $\sim \boldsymbol{L}(o1e)$.

These points determine the following three lines (fig. 37):

Def 4.1.1. $F = [o1]$, $E = [oe]$, $I = [o\boldsymbol{P}(o1e)]$.

We have the following obvious[2]

Th 4.1.2. *$o \in F, E, I$ and $\neq (F, E, I)$.*

The coordinate system and scalar algebra will depend on the *triple of parameters* $(o, 1, e)$ either directly or by means of F, E, and I. In particular, F will be an *axis* of the coordinate system and simultaneously the universe of the algebra. Its points will be called *scalars* and represented by the variables x, y, z and t, u, v. The fixed scalars o and 1 will play the role of zero and unit of the algebra.[3] The line E will be called *pseudo-axis* and its points *pseudo-scalars*.[4] The following function ε connects scalars with pseudo-scalars:

Def 4.1.3. $o^\varepsilon = o$ *and* $x \neq o \Rightarrow x^\varepsilon \in E \wedge \boldsymbol{P}(oxx^\varepsilon) \in I$.

We have the following obvious

Th 4.1.4. (i) *The function $\varepsilon : F \to E$ is a bijection.*
(ii) $1^\varepsilon = e$.

The function $\mathbf{L}_F$ assigns to every scalar a certain line parallel to F (fig. 38):

Def 4.1.5. $\mathbf{L}_F(t) = \mathbf{L}_{\|F}(t^\varepsilon)$.

Obviously

Th 4.1.6. (i) *The function $\mathbf{L}_F : F \to \{P : P \| F\}$ is a bijection*
(ii) $\mathbf{L}_F(o) = F$.

---

[2] In this section we shall in general omit the proofs, since they are very simple.

[3] Points are denoted by small Latin letters. Here, exceptionally, a point is denoted by a numeral because of its role in the scalar algebra.

[4] In fact, the line F will play the role of the second axis of the coordinate system, and therefore the line E will play the role of the first one.

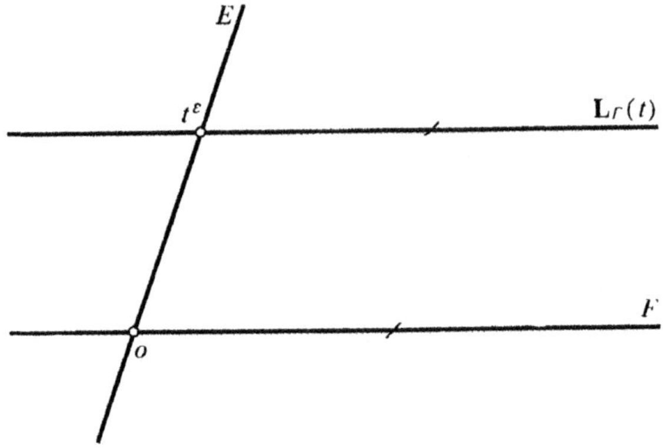

Fig. 38

4.2. CONSTRUCTION OF THE CARTESIAN COORDINATE SYSTEM

The function $\boldsymbol{p}$ assigns points of the plane $\mathfrak{A}$ to ordered pairs of scalars (fig. 39):

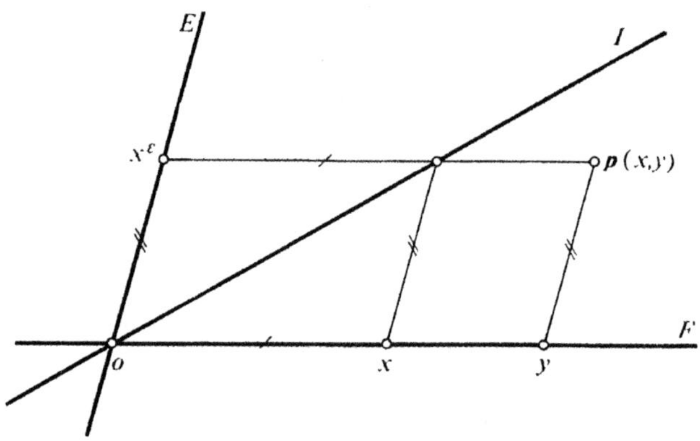

Fig. 39

Def 4.2.0. $\boldsymbol{p}(x, o) = x^\varepsilon$ and $\boldsymbol{p}(o, y) = y$,
$x, y \neq o \Rightarrow \boldsymbol{p}(x, y) = \boldsymbol{P}(ox^\varepsilon y)$.

Obviously[5]

Th 4.2.1. *The function* $\boldsymbol{p}: F^2 \to \boldsymbol{S}$ *is a bijection.*

Let Π be the inverse of $\boldsymbol{p}$:

Def 4.2.2. $\Pi(p) = (x, y) \Leftrightarrow \boldsymbol{p}(x, y) = p$.

[5] See footnote 2.

Each of two functions $\boldsymbol{p}$ and Π will be called the *Cartesian coordinate system with basis* $(o, 1, e)$. Let us set

Def 4.2.3. $\Pi(p) = (x_p, y_p)$.

The scalars x_p and y_p will be called the *first* and *second coordinates* of $\boldsymbol{p}$.

The following theorem connects points with their coordinates by means of the relations of parallelogram and trapezoid (figs. 40 and 41). It will be useful in the sequel.

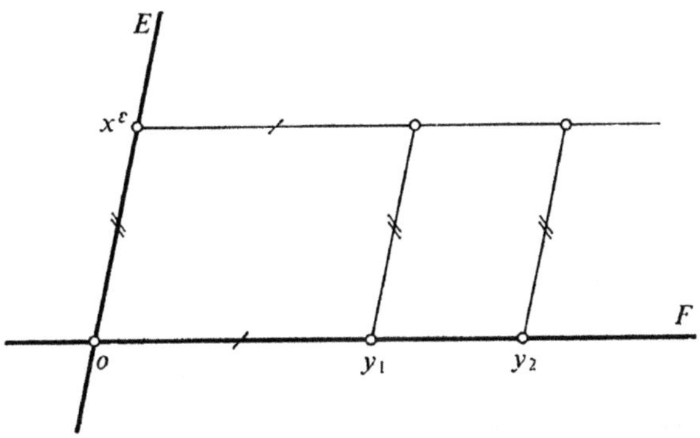

Fig. 40

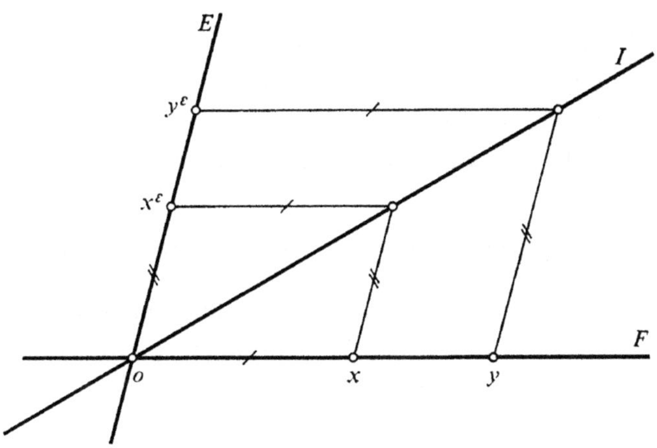

Fig. 41

Th 4.2.4. (i) $x \neq o \wedge y_1 \neq y_2 \Rightarrow \mathbf{P}(\boldsymbol{p}(x, y_1)\boldsymbol{p}(x, y_2) y_1 y_2)$,
(ii) $x, y \neq o \wedge x \neq y \Rightarrow \boldsymbol{T}_o(x\, y\, \boldsymbol{p}(x, x)\boldsymbol{p}(y, y))$.

The function **L** assigns to ordered pairs of scalars some lines non-parallel to F (fig. 42):

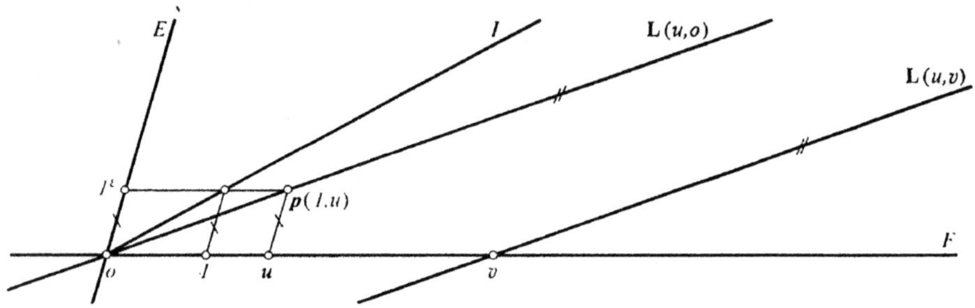

Fig. 42

Def 4.2.5. $\mathbf{L}(u, o) = [o\, \boldsymbol{p}(1, u)]$,
$\mathbf{L}(u, v) = \mathbf{L}_{||\mathbf{L}(u,\, o)}(v)$.

Obviously

Th 4.2.6. (i) *The function* $\mathbf{L}$: $F^2 \to \{P: \sim P||F\}$ *is a bijection*
(ii) $\mathbf{L}(u, v)||\mathbf{L}(u'v') \Leftrightarrow u = u'$.

By 4.1.6 (i) and 4.2.6 (i), every line of the plane is either of the form $\mathbf{L}_F(t)$ or $\mathbf{L}(u, v)$. Some of the lines can be already (without a scalar algebra) described by means of linear equation:

Th 4.2.7. (i) $\mathbf{L}_F(t) = \{\boldsymbol{p}(x, y): x = t\}$
and, in particular, $F = \mathbf{L}_F(o) = \{\boldsymbol{p}(x, y): x = o\}$.
(ii) $\mathbf{L}(o, v) = \{\boldsymbol{p}(x, y): v = y\}$
and, in particular, $E = \mathbf{L}(o, o) = \{\boldsymbol{p}(x, y): o = y\}$.
(iii) $I = \mathbf{L}(1, o) = \{\boldsymbol{p}(x, y): x = y\}$.

4.3. CONSTRUCTION OF THE DOUBLE LOOP

By means of the coordinate system $\boldsymbol{p}$ we define the sum $x+v$ (fig. 43) of any two scalars x and v, and the product $u \cdot x$ (fig. 44) of any two scalars u and x:[6]

[6] Such a choice of variables and their order will be justified in the sequel (compare the equation of a line).

Def 4.3.0. (i) $x+o = o+x = x.$
(ii) $x, v \neq o \Rightarrow \boldsymbol{P}(\boldsymbol{p}(o, o)\boldsymbol{p}(x, x)\boldsymbol{p}(o, v)\boldsymbol{p}(x, x+v)).$

Def 4.3.1. (i) $x \cdot 1 = 1 \cdot x = x.$
(ii) $x \cdot o = o \cdot x = o.$
(iii) $u, x \neq o, 1 \Rightarrow \boldsymbol{T}_o(\boldsymbol{p}(1, 1)\boldsymbol{p}(x, x)\boldsymbol{p}(1, u)\boldsymbol{p}(x, u \cdot x)).$

Thus $\boldsymbol{L}(o\boldsymbol{p}(1, u)\boldsymbol{p}(x, u \cdot x))$, which yields

Th 4.3.2. $u, x \neq o, 1 \Rightarrow \boldsymbol{T}_o(\boldsymbol{p}(1, u)\, \boldsymbol{p}(x, u \cdot x)\; u\;\; u \cdot x).$

Definitions 4.3.0 and 4.3.1 give at once the linear equations of the lines $\mathbf{L}(1, v)$ and $\mathbf{L}(u, o)$:

Th 4.3.3. (i) $\mathbf{L}(1, v) = \{\boldsymbol{p}(x, y)\colon x+v = y\}.$
(ii) $\mathbf{L}(u, o) = \{\boldsymbol{p}(x, y)\colon u \cdot x = y\}.$

Let us consider the algebras

$$\mathfrak{F}_+(o, 1, e) = (F, o, +) \text{ and } \mathfrak{F}_\cdot(o, 1, e) = (F, o, 1, \cdot).$$

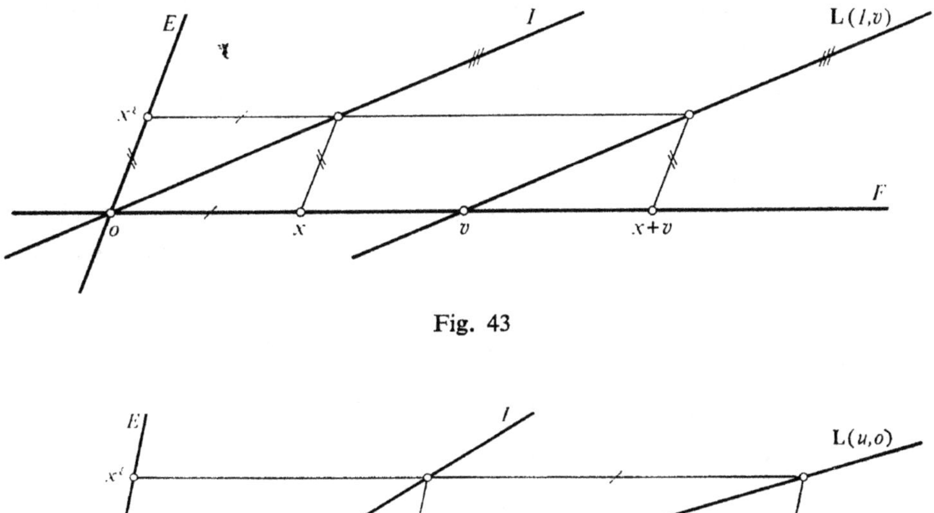

Fig. 43

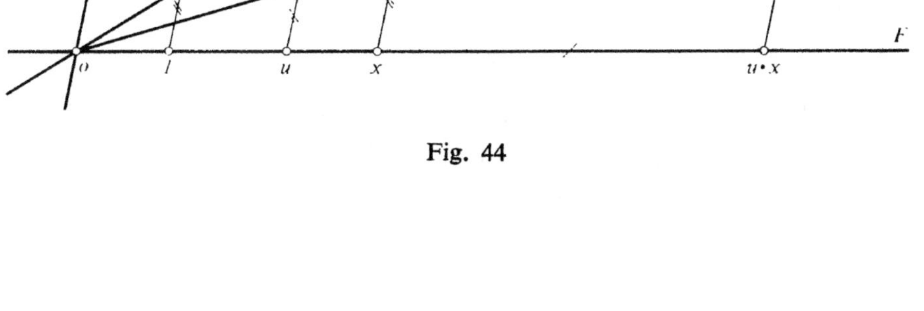

Fig. 44

It is easy to check that $\mathfrak{F}_+(o, 1, e)$ is a loop (additive) and $\mathfrak{F}_.(o, 1, e)$ is a multiplicative loop with zero. Thus we have

Th 4.3.4. *The algebra*

$$\mathfrak{F}(o, 1, e) = (F, o, 1, +, \cdot)$$

is a double loop.

Let us consider the bijections σ^v and ${}^u\psi$ of the double loop $\mathfrak{F}(o, 1, e)$ (see 0.6.1). As a direct consequence of 4.3.0 and 4.3.1 we get

Th 4.3.5. *Each of the functions* σ^v and ${}^u\psi$ *for* $u \neq o$ *can be represented as a composition of three parallel projections*:

(i) $\sigma^v\colon F \underset{\|E}{\rightarrow} I \underset{\|F}{\rightarrow} \mathbf{L}(1, v) \underset{\|E}{\rightarrow} F,$

(ii) ${}^u\psi\colon F \underset{\|E}{\rightarrow} I \underset{\|F}{\rightarrow} \mathbf{L}(u, o) \underset{\|E}{\rightarrow} F.$

Assume now that $\mathfrak{A}$ is a translation plane. Take two arbitrary scalars $x, v \in F$ and let $\alpha = \tau(ov)$. Then $o^\alpha = v$, and, by 4.3.0 (ii) and 4.2.4 (i), for $x, v \neq o$

$$o^\alpha = v \Rightarrow p(x, x)^\alpha = p(x, x+v) \Rightarrow x^\alpha = x+v,$$

i.e.

$$x+v = x^{\tau(ov)}.$$

By 4.3.0 (i) this formula remains valid for $x = o$ and for $v = o$. Hence, the operation $+$ coincides with the operation $+_o$ on F (see 3.4.3); this together with 3.4.6 results in

Th 4.3.6. *The loop* $\mathfrak{F}_+(o, 1, e)$ *depends only on the points* o *and* 1
des
and is an additive group of the line F, with o as zero.

Thus we define $\mathfrak{F}_+(o, 1)$ by the formula

$$\mathfrak{F}_+(o, 1) = \mathfrak{F}_+(o, 1, e).$$

Let us pass to the double loop $\mathfrak{F}(o, 1, e)$. By 4.3.5 (ii) and 3.4.8, for fixed $u \neq o$ the operation ${}^u\psi$ is an automorphism of the additive group $\mathfrak{F}_+(o, 1)$, i.e.

$$u \cdot (x+y) = u \cdot x + u \cdot y.$$

By 4.3.1 (ii) and 4.3.0 (i) this formula remains valid for $u = o$. Thus, in the double loop $\mathfrak{F}(o, 1, e)$ the law of left distributivity of the multiplication with respect to addition is valid. This together with 4.3.6 (because of the commutativity of the additive group of F) gives

Th 4.3.7. *The double loop $\mathfrak{F}(o, 1, e)$ is a left quasi-field.*
des

Let us notice that this quasi-field essentially depends (by means of multiplication) on the whole triple of parameters $(o, 1, e)$.

Assume, in turn, that $\mathfrak{A}$ is a homothety plane. Take two arbitrary scalars $u, x \in F - \{o\}$ and let $\varkappa = \iota_o(1x)$. Then $o^\varkappa = o$ and $1^\varkappa = x$, and, by 4.2.4 (ii), 4.3.1 (iii), and 4.3.2,

$$o^\varkappa = o \wedge 1^\varkappa = x \mapsto p(1, 1)^\varkappa = p(x, x) \mapsto p(1, u)^\varkappa = p(x, u \cdot x)$$
$$\mapsto u^\varkappa = u \cdot x,$$

i.e.

$$u \cdot x = u^{\iota_o(1x)}.$$

By 4.3.1 (i) this formula remains valid for $u = 1$ and $x = 1$. Thus on the set $F - \{o\}$ the operation $\cdot$ coincides with $\cdot_o^1$ (see 3.5.3). This, together with 3.5.5, 3.5.6, and 4.3.1 (ii), results in

Th 4.3.8. *The loop $\mathfrak{F}.(o, 1, e)$ depends only on the points o and 1,*
Des
and is a multiplicative group of the line F, with o as zero and 1 as unit.

Thus, we define $\mathfrak{F}.(o, 1)$ by the formula

$$\mathfrak{F}.(o, 1) = \mathfrak{F}.(o, 1, e).$$

By 3.5.9, theorems 4.3.6 and 4.3.8 yield

Th 4.3.9. *The double loop $\mathfrak{F}(o, 1, e)$ depends only on the points*
Des
o and 1 and is a field of the line F with the points o and 1 as zero and unit.

Thus, we define $\mathfrak{F}(o, 1)$ by the formula

$$\mathfrak{F}(o, 1) = \mathfrak{F}(o, 1, e).$$

By 3.5.11, theorem 4.3.9 results in

Th 4.3.10. *The field $\mathfrak{F}(o, 1)$ is commutative.*
Papp

4.4. CONSTRUCTION OF THE TERNARY FIELD AND ALGEBRAIC DESCRIPTION OF THE PARALLELITY RELATION

We start with the linear equation of a line. It was established for all the lines $\mathbf{L}_F(t)$ (see 4.2.7 (i)) and for $\mathbf{L}(u, v)$ whenever $u = o$ or $u = 1$ or $v = o$ (see 4.2.7 (ii) and 4.3.3). The algebraic description of the remaining lines requires some preparation.

Let us assign to any scalar $u \neq o$ the addition operation $+^u$ (it suffices to define the sum $u \cdot x +^u v$ for arbitrary two scalars x and v):[7]

Def 4.4.0. *For arbitrary* $u \neq o$

(i) $u \cdot x +^u o = u \cdot x$ *and* $o +^u v = v$,

(ii) $x, v \neq o \Rightarrow \mathbf{P}(\mathbf{p}(o, o)\mathbf{p}(x, u \cdot x)\mathbf{p}(o, v)\mathbf{p}(x, u \cdot x +^u v))$.

Comparing definition 4.4.0 with 4.3.0, we obtain

Th 4.4.1. $+^1 = +$.

From 4.4.0, 4.2.7 (ii), and 4.3.3 (ii), we get at once (fig. 45)

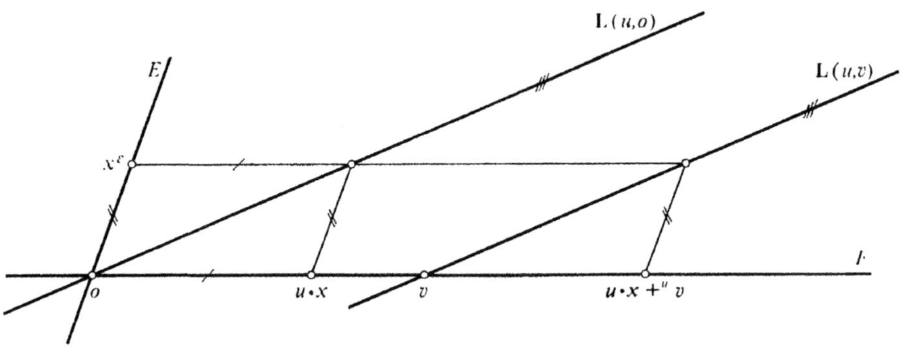

Fig. 45

Th 4.4.2. $\mathbf{L}(u, v) = \{\mathbf{p}(x, y): \; u \cdot x +^u v = y\}$.

Assume now that $\mathfrak{A}$ is a translation plane. Take arbitrary scalars $u, x, v \in F$ and let $\alpha = \tau(ov)$. Then $o^\alpha = v$ and, by 4.4.0 (ii) and 4.2.4 (i), for $u, x, v \neq o$

$$o^\alpha = v \mapsto \mathbf{p}(x, u \cdot x)^\alpha = \mathbf{p}(x, u \cdot x +^u v) \mapsto (u \cdot x)^\alpha = u \cdot x +^u v,$$

i.e.

$$u \cdot x +^u v = (u \cdot x)^{\tau(ov)}.$$

---

[7] Let us recall that the functions ${}^u\psi$ are bijections (see 0.6.1(ii)).

By 4.4.0 (i) this formula remains valid for $x = o$ and for $v = o$. Thus for arbitrary $u \neq o$ the operation $+^u$ coincides with the operation $+_o$ on the line F. As a result, $+^u = +^1$; this enables us to generalize 4.4.1:

Th 4.4.3. $+^u = +$ *for* $u \neq o$.
des

By 4.4.2 and 4.4.3 we have

Th 4.4.4. $\mathrm{L}(u, v) = \{p(x, y): u \cdot x + v = y\}$.
des

Theorems 4.2.7 (i) and 4.4.4 give an algebraic description of an arbitrary line of $\mathfrak{A}$. Hence, theorem 4.3.7 can be strenghtened and completed by the algebraic description of the parallelity relation (see 4.2.3):

Th 4.4.5. (i) *The double loop* $\mathfrak{F}(o, 1, e)$ *is a strong left quasi-field.*
des

(ii) $$ab\|cd \Leftrightarrow (x_a = x_b \wedge x_c = x_d \vee \exists u\, (u \cdot (x_a - x_b) = y_a - y_b \wedge u \cdot (x_c - x_d) = y_c - y_d)).$$

P r o o f. (i). By 4.2.6 (ii), if $u \neq u'$ then the lines $\mathrm{L}(u, v)$ and $\mathrm{L}(u', v')$ are not parallel and thus they have a point of intersection $p(x, y)$; by 4.4.4, its coordinate x satisfies the condition

$$u \cdot x + v = u' \cdot x + v'.$$

Hence the double loop $\mathfrak{F}(o, 1, e)$, which by 4.3.7 is a left quasi-field, satisfies the additional axiom 0.2.2str, and therefore it is a strong left quasi-field.

(ii). By 4.2.7 (i), 4.4.4, and 4.2.6 (ii), from theorem 2.3.5 it follows that

$$\begin{aligned} ab\|cd \Leftrightarrow x_a = x_b \wedge x_c = x_d \\ \vee \exists uvv' (u \cdot x_a + v = y_a \wedge u \cdot x_b + v = y_b \\ \wedge u \cdot x_c + v' = y_c \wedge u \cdot x_d + v' = y_d), \end{aligned}$$

whence, by simple calculation in left quasi-field, we get (ii). ∎

Let us observe that if, in particular, $\mathfrak{A}$ is a homothety plane (and thus, by 4.3.9, its scalar algebra is the field $\mathfrak{F}(o, 1)$), then by 0.6.4, the parallelity can be described algebraically in terms of the right Cartesian linear space $\mathfrak{V}^2(\mathfrak{F}(o, 1))$:

Th 4.4.6. $$ab\|cd \Leftrightarrow \exists uv \big((u, v) \neq (o, o) \wedge (\Pi(a) - \Pi(b)) \cdot u = (\Pi(c) - \Pi(d)) \cdot v\big).$$
Des

Theorem 4.4.5 together with 4.3.9, 4.3.10, and 4.4.6 completes the process of coordinatization of arbitrary weakly Desarguean plane, and in particular of any Desarguean or Pappian plane.

It remains to consider the most general case of an arbitrary parallelity plane $\mathfrak{A}$. Unfortunately, in this case, in the formula 4.4.2, the operation $+^u$ dependent by means of u on the direction of $\mathbf{L}(u, v)$, cannot be replaced by the operation $+$ of the double loop $\mathfrak{F}(o, 1, e)$. To obtain an algebraic description of the line $\mathbf{L}(u, v)$ it is necessary to change this scalar algebra, replacing its two binary operations $+$ and $\cdot$ by one ternary operation $\mathbf{T}$:

Def 4.4.7. (i) $\mathbf{T}(oxv) = v,$

(ii) $u \neq o \Rightarrow \mathbf{T}(uxv) = u \cdot x +^u v.$

Then

Th 4.4.8. $\mathbf{L}(u, v) = \{p(x, y) \colon \mathbf{T}(uxv) = y\}.$

Let us consider the algebra $\mathfrak{T}(o, 1, e) = (F, o, 1, \mathfrak{T})$. It is easy to verify in $\mathfrak{T}(o, 1, e)$ all the axioms for ternary fields. Indeed, from 4.1.0 it follows that the axiom 0.4.0 is satisfied; from 4.4.0(i), by 4.4.7(i), it follows that the axioms 0.4.1–0.4.4 are satisfied; from 2.3.9, 2.3.10, 1.2.4, 2.3.2, and 1.2.9 respectively, by 4.4.8 and 4.2.6(ii), it follows that the axioms 0.4.5–0.4.9 are satisfied. This, together with 4.4.8 and 4.2.6(ii), yields

Th 4.4.9. (i) *The algebra* $\mathfrak{T}(o, 1, e) = (F, o, 1, \mathbf{T})$ *is a ternary field.*

(ii) $$ab \| cd \Leftrightarrow x_a = x_b \wedge x_c = x_d \vee \exists uvv' \, (\mathbf{T}(ux_av) = y_a \wedge \mathbf{T}(ux_bv) = y_b \wedge \mathbf{T}(ux_cv') = y_c \wedge \mathbf{T}(ux_dv') = y_d).$$

Finally, let us establish the relationship between the ternary field $\mathfrak{T}(o, 1, e)$ and the double loop $\mathfrak{F}(o, 1, e)$. By 4.4.7 and 4.4.0(i), we get

Th 4.4.10. $x+v = \mathbf{T}(1xv)$ *and* $u \cdot x = \mathbf{T}(uxo).$

Thus $\mathfrak{F}(o, 1, e)$ is a loop of the ternary field $\mathfrak{T}(o, 1, e)$.

If $\mathfrak{A}$ is a translation plane, then by 4.4.3 from 4.4.7 we get

Th 4.4.11. $\mathbf{T}(uxv) = u \cdot x + v.$
des

Hence, on the translation plane, the loop $\mathfrak{F}(o, 1, e)$ is definitionally equivalent to the ternary field $\mathfrak{T}(o, 1, e)$.

In conclusion, replacement of the double loop by the ternary field did not change in fact the scalar algebra of the weakly Desarguean plane (and in particular of strongly Desarguean or Pappian plane); however, it enabled us to coordinatize an arbitrary parallelity plane. But such a solution is rather formal. Ternary field is a very weak algebra and it is difficult to calculate in it. The axiomatic method seems to be more suitable for arbitrary affine planes.

4.5. REPRESENTATION THEOREMS

We are now going to connect the classes of planes $\mathscr{A}$, des $\mathscr{A}$, Des $\mathscr{A}$, Papp $\mathscr{A}$ with the classes of algebras **TF**, **QF**, **F**, com **F** by means of representation theorems.

Consider an arbitrary parallelity plane $\mathfrak{A} = (S, \|)$, the system of parameters $(o, 1, e)$, the axis $F = [o1]$ and the Cartesian coordinate system Π with the basis $(o, 1, e)$ (see § 4.2). Let $\mathfrak{T} = (F^2, o, \cdot, \mathrm{T})$ be the ternary field with the basis $(o, 1, e)$ (see § 4.4) and let $\underline{\mathfrak{A}}(\mathfrak{T}) = (F^2, \|_{\mathfrak{T}})$ be the parallelity plane over $\mathfrak{T}$ (see § 4.0). We have

Th 4.5.0. $\mathfrak{A} \simeq \underline{\mathfrak{A}}(\mathfrak{T})$,
and the function Π is an isomorphism.

Proof: by 4.2.1, 4.2.2, 4.2.3 and 4.0.0, 4.4.9 (ii). ∎

Theorem 4.5.0 together with 4.0.1 yields the representation theorem for the class $\mathscr{A}$:

Theorem 4.5.1. $\mathscr{A} \underset{\mathrm{Iso}}{\subseteq} \{\underline{\mathfrak{A}}(\mathfrak{T}) : \mathfrak{T} \in \mathbf{TF}\} \subseteq \mathscr{A}$.

Let now $\mathfrak{A}$ be a translation plane. Let $\mathfrak{F} = (F, o, 1, +, \cdot)$ be a quasi-field with basis $(o, 1, e)$ (see 4.4.5 (i)) and let $\mathfrak{A}(\mathfrak{F}) = (F^2, \|_{\mathfrak{F}})$ be the parallelity plane over $\mathfrak{F}$ (see {§ 4.0). Then we have

Th 4.5.2. $\mathfrak{A} \simeq \underline{\mathfrak{A}}(\mathfrak{F})$,
des

and Π is an isomorphism.

Proof: by 4.2.1, 4.2.2, 4.2.3 and 4.0.2, 4.4.5 (ii). ∎

On the basis of 4.5.2 we get the representation theorem for the classes des $\mathscr{A}$, Des $\mathscr{A}$ and Papp $\mathscr{A}$:

Theorem 4.5.3. (i) $\operatorname{des}\mathscr{A} \underset{\text{Iso}}{\subseteq} \{\underline{\mathfrak{A}}(\mathfrak{F}) : \mathfrak{F} \in \mathbf{QF}\} \subseteq \operatorname{des}\mathscr{A}$.

(ii) $\operatorname{Des}\mathscr{A} \underset{\text{Iso}}{\subseteq} \{\underline{\mathfrak{A}}(\mathfrak{F}) : \mathfrak{F} \in \mathbf{F}\} \subseteq \operatorname{Des}\mathscr{A}$.

(iii) $\operatorname{Papp}\mathscr{A} \underset{\text{Iso}}{\subseteq} \{\underline{\mathfrak{A}}(\mathfrak{F}) : \mathfrak{F} \in \operatorname{com}\mathbf{F}\} \subseteq \operatorname{Papp}\mathscr{A}$.

Proof: (i) by 4.4.5 and 4.0.6 (i),
(ii) by 4.3.9 and 4.0.6 (ii),
(iii) by 4.3.10 and 4.0.6 (iii). ∎

We have

Theorem 4.5.4. *If* $\mathfrak{T}_1, \mathfrak{T}_2 \in \mathbf{TF}$, *then*

$$\mathfrak{T}_1 \simeq \mathfrak{T}_2 \Rightarrow \underline{\mathfrak{A}}(\mathfrak{T}_1) \simeq \underline{\mathfrak{A}}(\mathfrak{T}_2).$$

The natural question arises, when the converse implication is true.[8]

Let us treat $\mathfrak{A} = (S, ||) \in \mathscr{A}$ as a variable and introduce a function $\underline{\mathfrak{T}}$ assigning to any plane $\mathfrak{A}$ and a system of parameters $(o, 1, e)$, the ternary field

$$\mathfrak{T} = \underline{\mathfrak{T}}(\mathfrak{A}, (o, 1, e))$$

with basis $(o, 1, e)$.

Let $\mathfrak{A}_1, \mathfrak{A}_2 \in \mathscr{A}$ and let $(o_i, 1_i, e_i)$ be a system of parameters on $\mathfrak{A}_i$, $i = 1, 2$. We have the following obvious

Th 4.5.5. $(\mathfrak{A}_1, (o_1, 1_1, e_1)) \simeq (\mathfrak{A}_2, (o_2, 1_2, e_2))$
$\Rightarrow \underline{\mathfrak{T}}(\mathfrak{A}_1, (o_1, 1_1, e_1)) \simeq \underline{\mathfrak{T}}(\mathfrak{A}_2, (o_2, 1_2, e_2))$.

Theorem 4.5.0 may be reformulated in terms of the function $\underline{\mathfrak{T}}$ as follows:

Th 4.5.0.′ $\mathfrak{A} \simeq \underline{\mathfrak{A}}(\underline{\mathfrak{T}}(\mathfrak{A}, (o, 1, e)))$.

We shall prove the following

Th 4.5.6. $\mathfrak{T} \simeq \underline{\mathfrak{T}}(\underline{\mathfrak{A}}(\mathfrak{T}), ((0, 0), (0, 1), (1, 0)))$,

and the function f defined by the formula $f(x) = (0, x)$ is an isomorphism.

Proof. Let $\mathfrak{T} = (F, 0, 1, T) \in \mathbf{TF}$ and let $\mathfrak{A} = \underline{\mathfrak{A}}(\mathfrak{T})$. Then $\mathfrak{A} = (F^2, ||_{\mathfrak{T}})$ (see 4.0.0). So, we have the collinearity $L_{\mathfrak{T}}$, the parallelo-

[8] The remaining part of § 4.5 was written on the base of a sketch left by the author. The theorem 4.5.5 was lacking, while to the proof of 4.5.6 there was only a short hint (M.M).

gram relation $P_{\mathfrak{T}}$ and the trapezoid relation $(T_{(0,0)})_{\mathfrak{T}}$ defined in terms of $||_{\mathfrak{T}}$.

Choosing the system of parameters $((0, 0), (0, 1), (1, 0))$ in $\mathfrak{A}$, we fix thereby

$$\text{axis } F_{\mathfrak{A}} = [(0, 0)\ (0, 1)], \text{ pseudo-axis } E_{\mathfrak{A}} = [(0, 0)\ (1, 0)] \text{ and}$$

the line

$$I_{\mathfrak{A}} = [(0, 0)\mathbf{P}_{\mathfrak{T}}((0, 0)\ (0, 1)\ (1, 0))].$$

On the basis of 4.1.3, 4.1.1, 2.4.0, 4.0.0, and 0.4.4 we easily verify that

(1) $$(0, x)^{e} = (x, 0).$$

From 4.2.0 and (1), by 2.4.0 and 0.4.4, we get

(2) $$p_{\mathfrak{A}}((0, x), (0, y)) = (x, y).$$

Let $u, x \neq 0, 1$. From 4.3.1 (iii) and (2) it follows immediately that

$$(0, y) = (0, u) \underset{\mathfrak{A}}{\cdot} (0, x) \Leftrightarrow (T_{(0,0)})_{\mathfrak{T}}\ ((1, 1)(x, x)\ (1, u)\ (x, y)),$$

whence, by 2.4.0′, 4.0.0 and 0.4.1–0.4.3, we get

(3) $$(0, u) \underset{\mathfrak{A}}{\cdot} (0, x) = (0, \mathbf{T}(ux0)).$$

By 4.3.1 (i), (ii) and 0.4.1, 0.4.2, 0.4.4, the formula (3) remains valid for $u = 0$ as well as for $u = 1$ or $x = 0$ or $x = 1$.

Let now $u \neq 0$ and $x, v \neq 0$. From 4.4.0 (ii) together with (2) and (3) it follows that

$$(0, y) = (0, u) \underset{\mathfrak{A}}{\cdot} (0, x) \underset{\mathfrak{A}}{+}^{(0, u)} (0, v)$$

$$\Leftrightarrow P_{\mathfrak{T}}((0, 0)\ (x, \mathbf{T}(ux0))\ (0, v)\ (x, y)),$$

whence by 2.4.0, 4.0.0 and 0.4.2, 0.4.3, we obtain

(4) $$(0, u) \underset{\mathfrak{A}}{\cdot} (0, x) \underset{\mathfrak{A}}{+}^{(0, u)} (0, v) = (0, \mathbf{T}(uxv)).$$

By 4.4.0 (i), (3), and 0.4.3, the formula (4) remains valid for $x = 0$ or $v = 0$.

According to 4.4.7, the ternary field $\mathfrak{T}(\mathfrak{A}, ((0, 0), (0, 1), (1, 0)))$ is of the form

$$\mathfrak{T}(\mathfrak{A}, ((0, 0), (0, 1), (1, 0))) = (F_{\mathfrak{A}}, (0, 0), (0, 1), \mathbf{T}_{\mathfrak{A}}),$$

where

$$(5)\qquad \begin{aligned} &\mathbf{T}_{\mathfrak{A}}((0,0)\,(0,x)\,(0,v)) = (0,v),\\ u \neq 0 \Rightarrow\ &\mathbf{T}_{\mathfrak{A}}((0,u)\,(0,x)\,(0,v)) = (0,u) \underset{\mathfrak{A}}{\cdot} (0,x) \underset{\mathfrak{A}}{+}{}^{(0,u)} (0,v).\end{aligned}$$

From (4) and (5) with the help of 0.4.4 we get the following relationship between the operation $\mathbf{T}_{\mathfrak{A}}$ and the original operation $\mathbf{T}$:

$$(6)\qquad \mathbf{T}_{\mathfrak{A}}((0,u)\,(0,x)\,(0,v)) = (0, \mathbf{T}(uxv)).$$

Let $x \in F$. The formula

$$f(x) = (0, x)$$

defines a bijection $f\colon F \to F_{\mathfrak{A}}$. By (6), the bijection f is an isomorphism of the structures $\mathfrak{T}$ and $\underline{\mathfrak{T}}(\underline{\mathfrak{A}}(\mathfrak{T}), ((0,0), (0,1), (1,0)))$. ∎

Theorem 4.5.6 yields a partial answer to the question of the implication converse to that established in 4.5.4.

Theorem 4.5.7. *For every two* $\mathfrak{F}_1, \mathfrak{F}_2 \in \mathbf{F}$

$$\underline{\mathfrak{A}}(\mathfrak{F}_1) \simeq \underline{\mathfrak{A}}(\mathfrak{F}_2) \Rightarrow \mathfrak{F}_1 \simeq \mathfrak{F}_2.$$

Proof. Consider two isomorphic structures

$$\mathfrak{A}_i = \underline{\mathfrak{A}}(\mathfrak{F}_i) = (F_i^2, \|_{\mathfrak{F}_i}) \quad \text{for} \quad i = 1, 2.$$

Let $h\colon F_1^2 \to F_2^2$ be an isomorphism. Then h carries the system of parameters $((0_1, 0_1), (0_1, 1_1), (1_1, 0_1))$ of $\mathfrak{A}_1$ onto a system of parameters of $\mathfrak{A}_2$. Thus, by 4.5.5 and 3.5.10, we get

$$(1)\qquad \begin{aligned} &\underline{\mathfrak{F}}(\mathfrak{A}_1, ((0_1,0_1), (0_1,1_1), (1_1,0_1)))\\ &\simeq \underline{\mathfrak{F}}(\mathfrak{A}_2, ((0_2,0_2), (0_2,1_2), (1_2,0_2))).\end{aligned}$$

From 4.5.6 and (1) it follows at once that

$$\mathfrak{F}_1 \simeq \mathfrak{F}_2. \ ∎$$

Let us notice that for arbitrary $\mathfrak{F}_1, \mathfrak{F}_2 \in \mathbf{QF}$

$$\underline{\mathfrak{A}}(\mathfrak{F}_1) \simeq \underline{\mathfrak{A}}(\mathfrak{F}_2) \Rightarrow (\mathfrak{F}_1)_+ \simeq (\mathfrak{F}_2)_+.$$

Indeed, it suffices to base the proof of 4.5.7 upon 3.4.9 instead of 3.5.10. However, there exist $\mathfrak{F}_1, \mathfrak{F}_2 \in \mathbf{QF}$ such, that $\underline{\mathfrak{A}}(\mathfrak{F}_1) \simeq \underline{\mathfrak{A}}(\mathfrak{F}_2)$ though $\mathfrak{F}_1 \neq \mathfrak{F}_2$.[9]

---

[9] See Pickert [75], footnote 1, p. 97.

We are going to present one more reason for departing from the traditional treatment of affine plane as a collinearity structure (or as an incidence structure). Since we shall deal with two classes of structures, we distinguish them by indices. Let now $\mathscr{A}_{||}$ be the class of parallelity planes and $\mathscr{A}_L$ the class of affine planes treated as collinearity structures.

The set of sentences AX 2.2 describing the class $\mathscr{A}_{||}$ is clearly of $\forall\exists$-character. By the Łoś-Suszko theorem (see Łoś–Suszko [57]), it follows that the class $\mathscr{A}_{||}$ is summable, i.e., for an arbitrary increasing sequence of its structures, the union of these structures also belongs to $\mathscr{A}_{||}$:

$$\text{if } \mathfrak{A}_k \in \mathscr{A}_{||} \text{ and } \mathfrak{A}_{k-1} \subseteq \mathfrak{A}_k \text{ for } k = 1, 2, \ldots, \text{ then } \bigcup \mathfrak{A}_k \in \mathscr{A}_{||}.$$

To find an axiom system for the class $\mathscr{A}_L$, we eliminate the term $||$ from the axioms 2.1.1 and 2.1.2 by means of the definitional axiom 2.1.0. Because of the form of 2.1.0 and 2.1.1, the obtained axiom system will be of $\forall\exists\forall$-character. Moreover, as will be proved below, the class $\mathscr{A}_L$ is not summable, and thus for this class no $\forall\exists$-axiom system exists. This is an additional argument for the choice of parallelity as a primitive notion.

On the other hand, $\mathscr{A}_{||}$ and $\mathscr{A}_L$ constitute an example of two definition ılly equivalent classes (see 2.2.16 and 2.2.17), one of which is summable while the other is not.

Theorem. *The class $\mathscr{A}_L$ is not summable.*

P r o o f.[10] Let $\mathscr{P}$ be the class of non-trivial collinearity structures which satisfy the additional axiom:

Ax proj. $\forall abcd\, \exists p\, \big(L(abp) \wedge L(cdp)\big)$.

Of course

(1) $\mathscr{P} \cap \mathscr{A}_L = \emptyset$.

The set of sentences AX $1.1 \cup \{\text{proj}\}$ describing the class $\mathscr{P}$ is of $\forall\exists$-character, thus, by the Łoś–Suszko theorem, the class $\mathscr{P}$ is summable.

Consider a sequence of commutative fields $\mathfrak{F}_k$, $k = 0, 1, 2, \ldots$, each one being a transcendental extension of the previous one:

(2) $\mathfrak{F}_k = \mathfrak{F}_{k-1}(t_k)$, with t_k being an element of F_k transcendental

[10] M.M.—based on a sketch.

over $\mathfrak{F}_{k-1}, k = 1, 2, \ldots$

Define a sequence of structures

$$\mathfrak{P}_k = (P_k, \hat{L}_k), \quad k = 0, 1, \ldots$$

as follows:

$$P_k = \{(x_0, x_1, x_2) \in F_k^3 - \{0\}: x_0 = 1 \vee (x_0 = 0 \wedge x_1 = 1) \vee (x_0 = x_1 = 0 \wedge x_2 = 1)\}$$

and, for arbitrary $a, b, c \in P_k$,

$$\hat{L}_k(abc) \Leftrightarrow \exists (t, u, v) \in F_k^3 - \{0\} \;\; t \cdot a + u \cdot b + v \cdot c = 0.$$

It is easy to check, that $\mathfrak{P}_k$ is isomorphic to the projective plane over $\mathfrak{F}_k$; hence

$$\mathfrak{P}_k \in \mathscr{P} \quad \text{for} \quad k = 0, 1, 2, \ldots$$

Evidently

$$\mathfrak{P}_{k-1} \subseteq \mathfrak{P}_k \quad \text{for} \quad k = 1, 2, \ldots,$$

whence

(3) $\bigcup \mathfrak{P}_k \in \mathscr{P}.$

On each plane $\mathfrak{P}_k$ let us choose the line M_k:

$$M_k = \{(x_0, x_1, x_2) \in P_k: \; x_0 + t_k \cdot x_1 + t_k^2 \cdot x_2 = 0\}.$$

The condition (2) yields

(4) $M_k \subseteq P_k - P_{k-1}.$

Let now

(5) $A_k = P_k - M_k, \quad L_k = \hat{L}_k | A_k,$

and let

$$\mathfrak{A}_k = (A_k, L_k).$$

We shall prove

(6) $\mathfrak{A}_k \in \mathscr{A}_L \quad \text{for} \quad k = 0, 1, \ldots,$

By 4.0.1 and 2.2.16, it suffices to show that

(7) $\mathfrak{A}_k \simeq (F_k^2, L_{\mathfrak{F}_k}).$

Let

$$N_k = \{(x_0, x_1, x_2) \in P_k \colon x_0 = 0\}$$

and

$$\boldsymbol{L}'_k = \hat{\boldsymbol{L}}_k | (P_k - N_k).$$

Clearly

(8) $$\mathfrak{A}_k \simeq (P_k - N_k, \boldsymbol{L}'_k).$$

On the other hand, by 0.6.3 and 0.6.4, the bijection $\varphi\colon P_k - N_k \to F_k^2$ defined by the formula

$$\varphi(x_0, x_1, x_2) = (x_1, x_2)$$

satisfies the condition

$$\boldsymbol{L}'_k(abc) \Leftrightarrow \boldsymbol{L}_{\mathfrak{F}_k}(\varphi(a)\varphi(b)\varphi(c));$$

therefore

(9) $$(P_k - N_k, \boldsymbol{L}'_k) \underset{\varphi}{\simeq} (F_k^2, \boldsymbol{L}_{\mathfrak{F}_k}).$$

From (8) and (9) we deduce (7), which yields (6).

By (4) and (5) we get

$$A_{k-1} \subseteq P_{k-1} \subseteq A_k \subseteq P_k$$

and

$$\boldsymbol{L}_{k-1} \subseteq \hat{\boldsymbol{L}}_{k-1} \subseteq \boldsymbol{L}_k \subseteq \hat{\boldsymbol{L}}_k;$$

thus

(10) $$\mathfrak{A}_{k-1} \subseteq \mathfrak{P}_{k-1} \subseteq \mathfrak{A}_k \subseteq \mathfrak{P}_k.$$

From (10) it follows that

(11) $$\bigcup \mathfrak{A}_k = \bigcup \mathfrak{P}_k.$$

By (1), (3), and (11)

$$\bigcup \mathfrak{A}_k \notin \mathscr{A}_L,$$

whence, by (6) and (10), the class $\mathscr{A}_L$ is not summable. ∎

PART II

CHAPTER 5

MIDPOINT ALGEBRAS

5.1. GROUPS WITH THE OPERATOR $\frac{1}{2}$

Let $\mathfrak{G} = (G, 0, +)$ be a commutative group satisfying two additional axioms:

Ax 5.1.1. $\forall a \exists x \;\; x+x = a.$

Ax 5.1.2. $a+a = 0 \Rightarrow a = 0.$

Using 5.1.2 we obtain immediately

Th 5.1.3. $x+x = x'+x' \Rightarrow x = x'.$

By 5.1.1 and 5.1.3, for an arbitrary $a \in G$ the equation $x+x = a$ has a unique solution in G. This enables us to define the function $\frac{1}{2}$: $G \to G$, which assigns to any element a its *half* $\frac{1}{2}a$:

Def 5.1.4. $\frac{1}{2}a+\frac{1}{2}a = a.$

From the axioms 0.1.1, 0.1.4, 0.1.7 combined with 5.1.3 we derive

Th 5.1.5. (i) $\frac{1}{2}0 = 0.$

(ii) $\frac{1}{2}(a+b) = \frac{1}{2}a+\frac{1}{2}b.$

(iii) $\frac{1}{2}(a+a) = a.$

(iv) $\frac{1}{2}a = \frac{1}{2}b \Rightarrow a = b.$

Hence $\frac{1}{2}$ is an automorphism of $\mathfrak{G}$. This automorphism will be referred to as an *operator*, and $\mathfrak{G}$ as a *group with the operator* $\frac{1}{2}$. The operator $\frac{1}{2}$ makes it possible to introduce the operation $\oplus$ of *arithmetic mean* in $\mathfrak{G}$:

Def 5.1.6. $a \oplus b = \frac{1}{2}(a+b)$.

Let us consider the structure $(G, \oplus)$. In the next section we shall describe it axiomatically, and so independently of its connections with groups.

5.2. MIDPOINT ALGEBRAS

A groupoid $(G, \oplus)$ will be called a *midpoint algebra* if it satisfies the following axioms:

Ax 5.2.1. $a \oplus a = a$ (*idempotency*).

Ax 5.2.2. $a \oplus b = b \oplus a$ (*commutativity*).

Ax 5.2.3. $(a \oplus b) \oplus (c \oplus d) = (a \oplus c) \oplus (b \oplus d)$ (*bi-commutativity*).

Ax 5.2.4. $\forall ab\, \exists x\; x \oplus a = b$.

The point $a \oplus b$ is then referred to as a *midpoint of the pair* (a, b) and $\oplus$ as *midpoint operation.*

Let us derive further properties of the operation $\oplus$:

Th 5.2.5. $(a \oplus b) \oplus c = (a \oplus c) \oplus (b \oplus c)$ (*self-distributivity*)

Proof. By 5.2.1 and 5.2.3

$$(a \oplus b) \oplus c = (a \oplus b) \oplus (c \oplus c) = (a \oplus c) \oplus (b \oplus c). \blacksquare$$

Th 5.2.6. $a \oplus b = a \Rightarrow a = b$.

Proof. Let

(1) $a \oplus b = a$.

By 5.2.4, for some $x \in G$

(2) $x \oplus a = b$.

By (1) and (2), from 5.2.1, 5.2.5, and 5.2.2 if follows that

$$b = b\oplus b = (x\oplus a)\oplus b = (x\oplus b)\oplus(a\oplus b) = (x\oplus b)\oplus a$$
$$= (x\oplus a)\oplus(b\oplus a) = b\oplus a = a. \blacksquare$$

Th 5.2.7. $x\oplus a = x'\oplus a \Rightarrow x = x'$.[1]

Proof. Let

(3) $x\oplus a = x'\oplus a.$

By 5.2.4, for some $y \in G$

(4) $y\oplus a = x.$

By (3) and (4), from 5.2.1, 5.2.2, and 5.2.3 if follows that

$$x = x\oplus x = x\oplus(y\oplus a) = (x\oplus y)\oplus(x\oplus a) = (x\oplus y)\oplus(x'\oplus a)$$
$$= (x\oplus x')\oplus(y\oplus a) = (x\oplus x')\oplus x = x\oplus(x\oplus x');$$

by 5.2.6

$$x = x\oplus(x\oplus x') \Rightarrow x = x\oplus x' \Rightarrow x = x'. \blacksquare$$

In view of 5.2.4 and 5.2.7, for arbitrary $a, b \in G$ the equation $x+a = b$ has exactly one solution in G. This allows to define the operation $\ominus$ inverse to $\oplus$:

Def 5.2.8. $(b\ominus a)\oplus a = b.$

Given a midpoint algebra $(G, \oplus)$, let us consider the structures

$(G, \oplus, a)$ for $a \in G$.

The distinguished element a of G will be referred to as *zero* of the structure $(G, \oplus, a)$, and the structure itself will be referred to as *midpoint algebra with zero*. When dealing with a definite midpoint algebra with zero, we shall denote this zero by "0".

Let $\mathfrak{G} = (G, \oplus, 0)$ be a midpoint algebra with the point $0 \in G$ as zero. We are now able to define *multiplication by* **2**, i.e. the function $\mathbf{2}\colon G \to G$ determined by

Def 5.2.9. $\mathbf{2}a\oplus 0 = \mathrm{a}.$

[1] See Hauschild and Rautenberg [74].

In other words, $\mathbf{2}a = a\ominus 0$. From 5.2.1, 5.2.5, and 5.2.7 we obtain immediately

Th 5.2.10. (i) $\mathbf{2}0 = 0$.
(ii) $\mathbf{2}(a\oplus b) = \mathbf{2}a\oplus\mathbf{2}b$.
(iii) $\mathbf{2}(a\oplus 0) = a$.
(iv) $\mathbf{2}a = \mathbf{2}b \Rightarrow a = b$.

Thus $\mathbf{2}$ is an automorphism of $\mathfrak{G}$. This automorphism will be referred to as an *operator*. The operator $\mathbf{2}$ enables us to introduce the addition operation $+$ in $\mathfrak{G}$.

Def 5.2.11. $a+b = \mathbf{2}(a\oplus b)$.

Let us consider the structure $(G, 0, +)$. In the next section we shall prove that it is a group with the operator $\frac{\mathbf{1}}{\mathbf{2}}$.

5.3. GROUPS WITH THE OPERATOR $\frac{1}{2}$ AS MIDPOINT ALGEBRAS WITH ZERO

Let $\mathbf{G}_{\left(\frac{1}{2}\right)}$ be the class of all the groups with the operator $\frac{\mathbf{1}}{\mathbf{2}}$, and let $\mathbf{M}_{(\text{zero})}$ be the class of all the midpoint algebras with zero. The following two theorems show that these two classes are definitionally equivalent.

Theorem 5.3.1. *If* $\mathfrak{G} = (G, 0, +) \in \mathbf{G}_{\left(\frac{1}{2}\right)}$ *and* $\oplus$ *is the arithmetic mean operation in* $\mathfrak{G}$ (see def 5.1.6), *then* $\Phi(\mathfrak{G}) = (G, \oplus, 0) \in \mathbf{M}_{(\text{zero})}$, *and the operation* $+$ *is the addition in* $\Phi(\mathfrak{G})$ (see def 5.2.11).

P r o o f. By simple calculation in the commutative group with the operator $\frac{\mathbf{1}}{\mathbf{2}}$ (using 5.1.5(ii), (iii), (iv)) we immediately check that the axioms 5.2.1–5.2.4 for midpoint algebras are satisfied in $(G, \oplus)$. Thus $\Phi(\mathfrak{G}) = (G, \oplus, 0) \in \mathbf{M}_{(\text{zero})}$. Moreover

$$0\oplus(a+b) = \frac{\mathbf{1}}{\mathbf{2}}(0+(a+b)) = \frac{\mathbf{1}}{\mathbf{2}}(a+b) = a\oplus b,$$

whence

$$a+b = \mathbf{2}(a\oplus b),$$

i.e. the operation $+$ is the addition in $\Phi(\mathfrak{G})$. ∎

Theorem 5.3.2. *If* $\mathfrak{G} = (G, \oplus, 0) \in \mathbf{M}_{(\text{zero})}$ *and* $+$ *is the addition in* $\mathfrak{G}$ (see def 5.2.11), *then* $\Psi(\mathfrak{G}) = (G, 0, +) \in \mathbf{G}_{\left(\frac{1}{2}\right)}$ *and* $\oplus$ *is the arithmetic mean operation in* $\Psi(\mathfrak{G})$ (see def 5.1.6).

Proof.[2] Calculating in the midpoint algebra with zero, we easily check (using 5.2.10(i)) that the axioms 0.1.1, 0.1.2weak, and 0.1.7 are satisfied in $(G, 0, +)$. The axiom 0.1.4 is also satisfied; indeed, by 0.1.1, 5.2.2, 5.2.3, and 5.2.10 (i)(ii),

$$\begin{aligned}(a+b)+c &= (a+b)+(c+0) = \mathbf{2}\big((a+b)\oplus(c+0)\big)\\ &= \mathbf{2}\big((a+0)\oplus(b+c)\big) = (a+0)+(b+c) = a+(b+c).\end{aligned}$$

By 5.2.1 and 5.2.10(i)(iv), we obtain 5.1.1 and 5.1.2. Thus $\Psi(\mathfrak{G}) \in \mathbf{G}_{\left(\frac{1}{2}\right)}$. Moreover

$$(a\oplus b)+(a\oplus b) = \mathbf{2}\big((a\oplus b)\oplus(a\oplus b)\big) = \mathbf{2}(a\oplus b) = a+b,$$

whence

$$a\oplus b = \frac{\mathbf{1}}{\mathbf{2}}(a+b),$$

i.e. $\oplus$ is the arithmetic mean in $\Psi(\mathfrak{G})$. ∎

Let us notice that, by 5.3.1 and 5.3.2, the definitions 5.1.6 and 5.2.11 determine two functors

$$\Phi\colon \mathbf{G}_{\left(\frac{1}{2}\right)} \to \mathbf{M}_{(\text{zero})} \quad \text{and} \quad \Psi\colon \mathbf{M}_{(\text{zero})} \to \mathbf{G}_{\left(\frac{1}{2}\right)}$$

such that

$$\Phi(G, 0, +) = (G, \oplus, 0) \Leftrightarrow \Psi(G, \oplus, 0) = (G, 0, +),$$

i.e. Φ and Ψ are mutually inverse.

For geometry, the notion of midpoint algebra is much more suitable than the notion of group. Whenever it is possible to construct a midpoint algebra, it suffices to choose a zero to obtain automatically (by means of Ψ) a commutative group.

---

[2] M.M.—based on a sketch.

We shall now briefly recall how the notion of vector may be introduced by means of midpoint. Let $(G, \oplus)$ be a midpoint algebra. For the elements of $G \times G$ (the so-called bound vectors) the notation $\overrightarrow{ab} = (a, b)$ is in common use. We define the congruence relation $\equiv$ on $G \times G$:

Def 5.4.0. $\overrightarrow{ab} \equiv \overrightarrow{cd} \Leftrightarrow a \oplus d = b \oplus c.$

By simple calculation in $(G, \oplus)$ we get

Th 5.4.1. $\overrightarrow{aa} = \overrightarrow{bb},$

Th 5.4.2. $\overrightarrow{aa} \equiv \overrightarrow{bc} \Rightarrow b = c,$

Th 5.4.3. $\overrightarrow{ab} \equiv \overrightarrow{ab},$

Th 5.4.4. $\forall abc\, \dot{\exists} d\ \overrightarrow{ab} \equiv \overrightarrow{cd},$ [3]

Th 5.4.5. $\overrightarrow{ab} \equiv \overrightarrow{pq} \wedge \overrightarrow{ab} \equiv \overrightarrow{rs} \Rightarrow \overrightarrow{pq} \equiv \overrightarrow{rs},$ [4]

Th 5.4.6. $\overrightarrow{ab} \equiv \overrightarrow{a'b'} \wedge \overrightarrow{bc} \equiv \overrightarrow{b'c'} \Rightarrow \overrightarrow{ac} \equiv \overrightarrow{a'c'}.$

In the proofs of the last two statements we "add" (in the sense of $\oplus$) corresponding sides of the assumed equalities, and then apply the bi-commutativity and the cancellation laws.

By 5.4.3 and 5.4.5, the relation $\equiv$ is an equivalence on the set $G \times G$. We define (free) vectors as the equivalence classes of $\equiv$:

Def 5.4.7. $[\overrightarrow{ab}] = \{\overrightarrow{pq} : \overrightarrow{ab} \equiv \overrightarrow{pq}\}.$

Let V be the set of all the vectors. In view of 5.4.1 and 5.4.2, all the degenerate pairs constitute one vector:

Def 5.4.8. $I = \{\overrightarrow{pq} : p = q\}.$

Hence

Th 5.4.9. $I = [\overrightarrow{aa}]$ for $a \in G.$

Let $\alpha \in V$; by 5.4.4, we have

Th 5.4.10. $\forall a\, \dot{\exists} b\ \alpha = [\overrightarrow{ab}].$

Theorem 5.4.10 enables us to define the sum of arbitrary two vectors as follows:

Def 5.4.11. $[\overrightarrow{ab}] + [\overrightarrow{bc}] = [\overrightarrow{ac}].$

---

[3] The symbol "$\dot{\exists} x$" reads: there is exactly one x (M.M).

[4] Compare theorem des_M in § 6.3 (M.M).

In view of 5.4.6, the sum is independent of the parameter a (the points b and c are uniquely determined by a).

Th 5.4.12. $[\overrightarrow{ab}] = -[\overrightarrow{ba}]$.

Proof: by 5.4.9 and 5.4.11. ∎

Th 5.4.13. $[\overrightarrow{a(a\oplus b')}]+[\overrightarrow{a(a\oplus b)}] = [\overrightarrow{ab}]$.

Proof: by 5.4.0, 5.4.11, and 5.2.1. ∎

Th 5.4.14. $\alpha+\alpha = I \Rightarrow \alpha = I$.

Proof: by 5.4.12 and 5.4.9. ∎

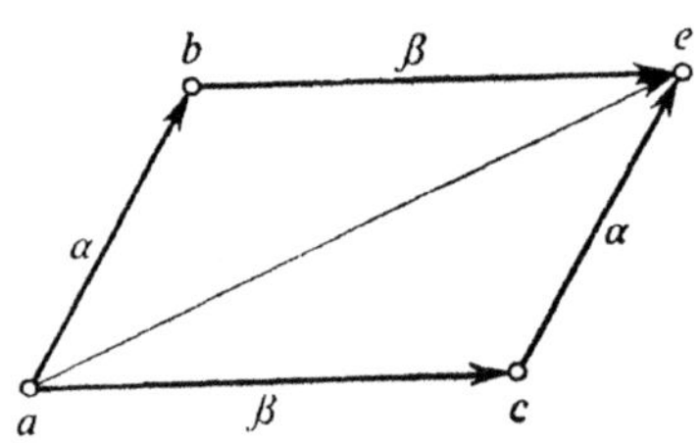

Fig. 46

Using 5.4.10 one can easily check that the structure $\mathfrak{V} = (V, I, +)$ is a group. This group is commutative; indeed, if (see fig. 46)

$$\alpha = [\overrightarrow{ab}],\ \beta = [\overrightarrow{bd}] = [\overrightarrow{ac}],\ \text{and}\ \alpha = [\overrightarrow{ce}],$$

then $a\oplus d = b\oplus c = a\oplus e$, whence $d = e$ and thus

$$\alpha+\beta = [\overrightarrow{ad}] = [\overrightarrow{ae}] = \beta+\alpha.$$

By 5.4.13 and 5.4.14 this is a group with the operator $\frac{1}{2}$. Thus, we have

Theorem 5.4.15. *The structure* $\mathfrak{V} = (V, I, +)$ *is a commutative group with the operator* $\frac{1}{2}$.

Let us now fix a point $o \in G$. We define the addition of points with o treated as a parameter:

Def 5.4.16. $a+b = c \Leftrightarrow [\overrightarrow{oa}]+[\overrightarrow{ob}] = [\overrightarrow{oc}]$.

Let $f_0(p) = [\overrightarrow{op}]$. The function f_0 is an isomorphism of the algebra $(G, o, +)$ onto the group $\mathfrak{V}$. Thus $(G, o, +)$ is a commutative group with the operator $\frac{1}{2}$. Let us observe that the same group of points was elementarily described in § 5.2. Indeed, from definitions 5.4.16 and 5.4.11 it follows that

$$a+b = c \Leftrightarrow [\overrightarrow{ob}] = [\overrightarrow{ac}],$$

i.e.

$$a+b = c \Leftrightarrow a\oplus b = o\oplus c.$$

Hence, by 5.2.9,

$$a+b = 2(a\oplus b),$$

but this is exactly the definition 5.2.11, which determines the operation + in § 5.2.

CHAPTER 6

TRANSLATION PLANES OF CHARACTERISTIC $\neq 2$

6.1. CHARACTERISTIC $\neq 2$

Let $\mathfrak{A} = (S, \|)$ be a translation plane. In view of 3.4.11, all the elements of Tr different from the identity are of the same order. This order may be either a natural (prime) number $n \geqslant 2$ or the infinity. The characteristic of $\mathfrak{A}$ is understood as this number n in the first case, and 0 in the second case.

If the characteristic of $\mathfrak{A}$ is equal 2 then, clearly, no element of Tr except the identity is divisible by 2. We are interested in planes of characteristic $\neq 2$. We are going to show that if $\mathfrak{A}$ is of characteristic $\neq 2$, then every element α of Tr is divisible by 2, i.e. $\alpha = \beta+\beta$ for some $\beta \in \mathrm{Tr}$. To this end let us pass to the additive group of the quasi-field of scalars. Consider an arbitrary translation α different from the identity, and choose a point o. Let

(1) $\qquad 1 = o^{\alpha}.$

We take the points o, 1, and an arbitrary point e satisfying the condition $\sim L(o1e)$, as parameters of coordinatization. Let $\mathfrak{F} = (F, o, 1, +, \cdot)$ be the quasi-field of scalars. Its additive group $\mathfrak{F}_+ = (F, o, +)$ is isomorphic to the subgroup Tr_F of Tr (see 3.4.6) and therefore it has no cyclic elements of order 2. In particular, $1+1 \neq o$. Thus, for some $u \in F$,

$$u \cdot (1+1) = 1;$$

so, calculating in $\mathfrak{F}$, we obtain

(2) $\qquad u+u = u \cdot 1+u \cdot 1 = u \cdot (1+1) = 1.$

We now define the translation β by

(3) $\qquad u = o^{\beta}.$

Of course, $\alpha, \beta \in \mathrm{Tr}_F$; by (1), (2), and (3), we get $\beta+\beta = \alpha$. So, we have

Th 6.1.1. *The following four conditions are equivalent:*
des

(i) $\mathfrak{A}$ *is of characteristic* $\neq 2$.
(ii) *Every element of* Tr *has order* $\neq 2$.
(iii) *Every element of* Tr *is divisible by* 2.
(iv) Tr *is a group with the operator* $\frac{1}{2}$.

In our proof we made an essential use of the embeddability of the group Tr_F in a left quasi-field.

The condition (iii) will be helpful in determining the characteristic of $\mathfrak{A}$ by means of an elementary axiom. Let us start with the following

Th 6.1.2. (I) $P(abcd) \wedge a^{\alpha} = d \wedge \alpha = \beta+\beta \Rightarrow \{a^{\beta}\} = [ad] \cap [bc]$.
des

(II) $P(abcd) \wedge a^{\alpha} = d \wedge \{a^{\beta}\} = [ad] \cap [bc] \Rightarrow \alpha = \beta+\beta$.

Proof. (I) Let $P(abcd)$ and $a^{\beta+\beta} = d$. Define the point p and the translations γ_1 and γ_2:

$$p = a^{\beta}, \quad \gamma_1 = \tau(bp), \quad \gamma_2 = \tau(pc) \quad \text{(fig. 47)}.$$

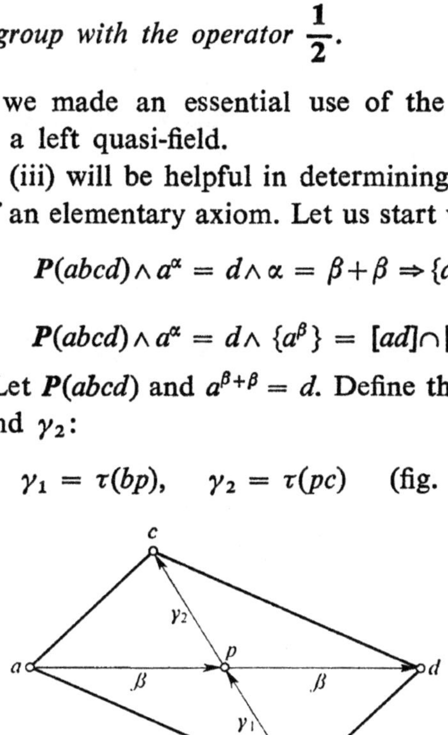

Fig. 47

Then $p^{\beta} = d$ and $\gamma_1+\beta = \beta+\gamma_2$, whence $\gamma_1 = \gamma_2$. From 2.5.9(ii) it follows that

$$\beta \in \mathrm{Tr}_{[ad]} \quad \text{and} \quad \gamma_1 \in \mathrm{Tr}_{[bc]};$$

thus, by 2.5.9(i),

$$p \in [ad] \quad \text{and} \quad p \in [bc].$$

(II) Let $P(abcd)$ and $\{p\} = [ad] \cap [bc]$. Then $\alpha = \tau(ad)$ and

$\beta = \tau(ap)$. Define the translations $\beta_1, \beta_2, \gamma_1, \gamma_2$:

$$\beta_1 = \beta = \tau(ap),\ \beta_2 = \tau(pd),\ \gamma_1 = \tau(bp),\ \gamma_2 = \tau(pc)$$

(fig. 48). Then $\beta_1 + \gamma_2 = \gamma_1 + \beta_2$, i.e.

$$\beta_1 - \beta_2 = \gamma_1 - \gamma_2;$$

since

$$\beta_1 - \beta_2 \in \mathsf{Tr}_{[ad]} \quad \text{and} \quad \gamma_1 - \gamma_2 \in \mathsf{Tr}_{[bc]},$$

by 2.5.8(ii) we get $\beta_1 = \beta_2$ and thus $\alpha = \beta + \beta$. ■

Applying now 6.1.2, we obtain

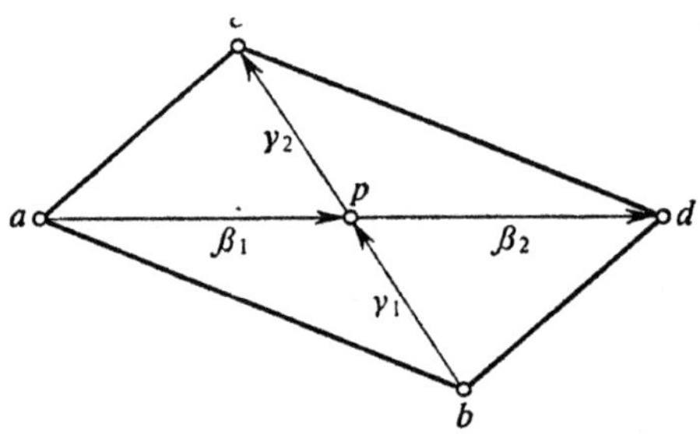

Fig. 48

Th 6.1.3. *The condition*
des

(v) $\quad P(abcd) \Rightarrow [ab] \cap [bc] \neq \emptyset,$

is equivalent to (iii).

P r o o f. (iii) $\Rightarrow$ (v): Assume (iii) and let $P(abcd)$. Define the translations α, β and the point p by the formulae:

$$\alpha = \tau(ad), \quad \alpha = \beta + \beta, \quad p = a^{\beta}.$$

Then, by 6.1.2(I), $p \in [ad] \cap [bc]$.

(v) $\Rightarrow$ (iii): Assume, conversely, (v) and let α be an arbitrary translation different from the identity. Let us choose a point a, and in succession points d, b, c, p and the translation β satisfying the conditions

$$d = a^{\alpha}, \quad \sim L(adb), \quad P(badc), \quad \{p\} = [ad] \cap (bc],$$
$$\beta = \tau(ap).$$

Then, by 6.1.2(II), we get $\alpha = \beta + \beta$. ■

The sentence (v) says that the diagonals of an arbitrary parallelogram have a point of intersection. In elementary language this sentence admits the form:

$$\sim L(abc) \wedge ab||cd \wedge ac||bd \Rightarrow \sim (ad||bc).$$

The following "positive" form seems to be more elegant:

Ax Ch. $ab||cd \wedge ac||bd \wedge ad||bc \Rightarrow L(abc)$.

Let us notice that, on the basis of the axiom system for parallelity planes, the axiom Ch is equivalent to Dim; hence, a translation plane of characteristic $\neq 2$ is just a translation plane with the collinearity structure of dimension 2 (see § 1). Any of the sentences Ch and Dim may be taken as an axiom ensuring to a plane $\mathfrak{A}$ the characteristic $\neq 2$. In our approach the axiom Ch seems more suitable.

Hence, the set of sentences

$$\text{AX } 2.2 \cup \{\text{des, Ch}\}$$

is an axiom system for translation planes of characteristic $\neq 2$.

By the argument used at the beginning of this section, we conclude that each of the conditions (ii) and (iii) can be replaced by an equivalent existential condition. In particular, the following condition (iii′) corresponds to (iii).

(iii)′ In the group Tr there is an element (different from the identity) divisible by 2.

Using 6.1.2 we easily show that (iii′) is equivalent to the following condition

(v′) $\exists abcd\ (P(abcd) \wedge [ad] \cap [bc] \neq \emptyset)$,

which leads to the existential form of the axiom Ch:

Ax Ch′. $\exists abcd\ (ab||cd \wedge ac||bd \wedge \sim (ad||bc))$.[1]

We omitted the condition $\sim L(abc)$ because it can be derived (on the basis of the axioms for parallelity planes) from the other conditions on a, b, c, d.

---

[1] Compare Fano's axiom for projective planes (see Pickert [75], p. 194)

6.2. GEOMETRIC MIDPOINTS

Let $\mathfrak{A} = (S, \|)$ be a translation plane of characteristic $\neq 2$.

By 6.1.1 the group Tr of $\mathfrak{A}$ is a group with the operator $\frac{1}{2}$. This enables us to define on S the operation $\hat{\oplus}$ of *geometric midpoint*:

Def 6.2.0. $\quad a\hat{\oplus}b = a^{\frac{1}{2}\tau(ab)}.$

Theorem 6.1.2(I) can be now reformulated as follows:

Th 6.2.1.[2] $\quad P(pabq) \Rightarrow \{a\hat{\oplus}b\} = [pq] \cap [ab].$
Ch

Theorem 6.2.1 describes the construction of $a\hat{\oplus}b$ for $a \neq b$: Consider a parallelogram with opposite vertices a, b; then $a\hat{\oplus}b$ is the point of intersection of $[ab]$ and the other diagonal of the parallelogram. Of course this construction is elementary.

Definition 6.2.0 yields

Th 6.2.2. $\quad a^{\beta+\beta} = b \Rightarrow a\hat{\oplus}b = a^{\beta}.$
Ch

From 6.2.2, calculating in the translation group, we derive basic properties of the operation $\hat{\oplus}$:

Th 6.2.3. $\quad a\hat{\oplus}a = a.$
Ch

P r o o f. In this case, the translation β coincides with the identity transformation. ∎

Th 6.2.4. $\quad a\hat{\oplus}b = b\hat{\oplus}a.$
Ch

P r o o f. Let

$$a^{\beta+\beta} = b.$$

Then

$$a = b^{(-\beta)+(-\beta)}$$

[2] The term "Th / Ch" indicates the translation planes of characteristic $\neq 2$, to which the theorem refers (compare footnote 17 in § 3.4).

and

$$a\hat{\oplus}b = a^{\beta} = b^{-\beta} = b\hat{\oplus}a. \blacksquare$$

Lem 6.2.5. $a\hat{\oplus}b = a^{\gamma}\hat{\oplus}b^{-\gamma}$ *for every* $\gamma \in \mathsf{Tr}$.
Ch

Proof. Let

$$a^{\beta+\beta} = b;$$

then

$$a^{\gamma+(\beta-\gamma)+(\beta-\gamma)} = b^{-\gamma},$$

which implies

$$a^{\gamma}\hat{\oplus}b^{-\gamma} = a^{\gamma+(\beta-\gamma)} = a^{\beta} = a\hat{\oplus}b. \blacksquare$$

Th 6.2.6. $(a\hat{\oplus}b)\hat{\oplus}(c\hat{\oplus}d) = (a\hat{\oplus}c)\hat{\oplus}(b\hat{\oplus}d)$.
Ch

Proof. Let

$$a^{\alpha+\alpha} = b, \quad c^{\beta+\beta} = d, \quad a^{\gamma+\gamma} = c, \quad b^{\delta+\delta} = d;$$

then

$$\alpha+\delta = \beta+\gamma.$$

We have to prove that

$$a^{\alpha}\hat{\oplus}c^{\beta} = a^{\gamma}\hat{\oplus}b^{\delta}.$$

Indeed, by 6.2.5,

$$a^{\alpha}\hat{\oplus}c^{\beta} = a^{\alpha}\hat{\oplus}d^{-\beta} = a^{\alpha+(\gamma-\alpha)}\hat{\oplus}d^{-\beta-(\gamma-\alpha)} = a^{\gamma}\hat{\oplus}d^{-\delta} = a^{\gamma}\hat{\oplus}b^{\delta}. \blacksquare$$

Th 6.2.7. $\forall ac \dot{\exists} b\, a\hat{\oplus}b = c$.
Ch

Proof. If $a^{\beta} = c$, then $b = a^{\beta+\beta}$. $\blacksquare$

Evidently, Theorem 6.2.7 is the conjunction of the following two statements:

Th 6.2.8. $\forall ac\, \exists b\, a\hat{\oplus}b = c$,
Ch

and

Th 6.2.9. $a\hat{\oplus}b = a\hat{\oplus}b' \Rightarrow b = b'$.
Ch

Theorems 6.2.3, 6.2.4, 6.2.6, and 6.2.8 can be expressed as

Theorem 6.2.10. *For an arbitrary translation plane* $\mathfrak{A} = (S, ||)$ *of characteristic* $\neq 2$, *the structure* $(S, \hat{\oplus})$ *is a midpoint algebra.*

The following theorem establishes the basic relationship between the operation $\oplus$ and parallelity:

Th 6.2.11. $a\hat{\oplus}b = c\hat{\oplus}d \Rightarrow ac||bd.$
Ch

Proof. If $a^{\gamma} = c$, then, by 6.2.5 and 6.2.9, $d^{\gamma} = b$. ∎

In view of 6.2.10, Theorems 5.2.5 and 6.2.11 yield

Th 6.2.12. $ab||(a\hat{\oplus}c)(b\hat{\oplus}c).$
Ch

6.3. MIDPOINT PLANES

The following two properties of Tr distinguish the translation planes of characteristic $\neq 2$ from among all the parallelity planes:

(I) Tr is transitive

(II) Tr is a group with the operator $\frac{1}{2}$.

The elementary counterparts of these two properties are the law des and the axiom Ch, respectively. The same class of planes can be also determined by means of midpoint algebras:

Theerem 6.3.0. *Let* $\mathfrak{A} = (S, ||)$ *be a parallelity plane. If there is an operation* $\oplus: S+S \to S$ *satisfying the conditions*:

(i) *The structure* $(S, \oplus)$ *is a midpoint algebra,*

(ii) $a\oplus b = c\oplus d \Rightarrow ac||bd$ *for* $a, b, c, d \in S$,

then $\mathfrak{A}$ *is a translation plane of characteristic* $\neq 2$, *and the operation* $\oplus$ *coincides with the operation* $\hat{\oplus}$ *of geometric midpoint.*

Proof. Assume there is an operation $\oplus$ on S satisfying (i) and (ii). Let us show that the group of vectors $\mathfrak{B} = (V, I, +)$ of the algebra $(S, \oplus)$ (see § 5.4) coincides with the translation group Tr of $\mathfrak{A}$. By 5.4.10, any vector α is a transformation, $\alpha: S \to S$, and consists of the pairs $\overrightarrow{pp^{\alpha}}$ for $p \in S$. All these pairs are congruent, $\overrightarrow{pp^{\alpha}} \equiv \overrightarrow{qq^{\alpha}}$, i.e. $p\oplus q^{\alpha} = q\oplus p^{\alpha}$; thus by (ii), $pq||p^{\alpha}q^{\alpha}$ for arbitrary $p, q \in S$. If $p^{\alpha} = p$, then

$\alpha = I$; thus α is a translation, and therefore

$$V \subseteq \mathsf{Tr}.$$

By 5.4.11, the addition of vectors coincides with the composition of translations. Thus $\mathfrak{V}$ is a subgroup of Tr; moreover, it is a transitive subgroup, because for every $p, q \in S$ the vector $\alpha = [\overrightarrow{pq}]$ carries p on q. Hence $\mathfrak{V}$ coincides with the whole group Tr, i.e.

$$\mathfrak{V} = \mathsf{Tr}.$$

By this equality, Tr is transitive, and, by 5.4.15, it is a group with the operator $\frac{1}{2}$. Consequently, $\mathfrak{A}$ is a translation plane of characteristic $\neq 2$.

It remains to show that the operation $\oplus$ coincides with $\hat{\oplus}$. Let

$$\alpha = \tau(ab) \quad \text{and} \quad \beta = \frac{1}{2}\alpha.$$

Using the vector notation, we have

$$\alpha = [\overrightarrow{ab}],$$

whence, by 5.4.13,

$$\beta = [\overrightarrow{a(a\oplus b)}],$$

and therefore

$$a\oplus b = a^\beta = a\hat{\oplus}b. \blacksquare$$

Let $\mathfrak{A} = (S, \|, \oplus)$ be an arbitrary structure satisfying the following three conditions:

(o) The structure $(S, \|)$ is a parallelity plane,

(i) The structure $(S, \oplus)$ is a midpoint algebra,

(ii) $a\oplus b = c\oplus d \Rightarrow ac\|bd$ for $a, b, c, d \in S$.

According to Theorem 6.3.0, the structure $(S, \|)$ is then a translation plane of characteristic $\neq 2$, and for any two points $a, b \in S$ the point $a\oplus b$ is the geometric midpoint of the pair (a, b), defined by 6.2.0 or by the (equivalent) conjunction of two elementary sentences 6.2.1 and 6.2.2.[3] Hence, the structure $\mathfrak{A}$ is a definitional expansion of $(S, \|)$ and

---

[3] Of course, the sentence 6.2.1, which describes $a\hat{\oplus}b$ for $a \neq b$, can be easily expressed in terms of $\|$ directly.

the only difference between $\mathfrak{A}$ and $(\boldsymbol{S}, ||)$ is that in the former structure the operation $\oplus$ is given as a primitive notion. The structure $\mathfrak{A} = (\boldsymbol{S}, ||, \oplus)$ satisfying (o), (i), and (ii) will be referred to as a *midpoint plane*. So, in fact, there is no essential difference between the midpoint planes and the translation planes of characteristic $\neq 2$. However, the axiom systems are different. To obtain an axiom system for the midpoint planes from the axiom system for translation planes of characteristic $\neq 2$, we preserve AX 2.2 and replace the axioms des and Ch by the axioms of midpoint algebra, 5.2.1–5.2.4, and by the axiom

Ax 6.3.1. $a \oplus b = c \oplus d \Rightarrow ac||bd,$

connecting $\oplus$ with $||$. Hence the set of sentences

$$\text{AX } 6.3 = \text{AX } 2.2 \cup \{5.2.1\text{–}5.2.4\} \cup \{6.3.1\}$$

is an axiom system for midpoint planes.

The axiom 6.3.1 reads as follows: every reflection is a dilatation.[4] As a particular case of 6.3.1, we have

Th 6.3.2.[5] $L(ab(a \oplus b))$.
Mid

Proof. $(a \oplus b) \oplus (a \oplus b) = a \oplus b \mathrel{|}\!\Rightarrow (a \oplus b)a||(a \oplus b)b$
$\mathrel{|}\!\Rightarrow L(ab(a \oplus b))$. ■

This new axiomatic approach has many advantages. Indeed, the midpoint algebra is now given directly, and, as will be shown, it does duty for the translation group; when calculating in that algebra, we deal with points, i.e. with elements of the universe itself. On the other hand, changing the notation for midpoint plane from $(\boldsymbol{S}, ||, \oplus)$ into $(\boldsymbol{S}, \oplus, ||)$ and changing respectively the order of axioms, we observe that the axiom system for the planes is adequate to that for the quasi-fields of scalars. This results in a very short and simple process of coordinatization. Namely, given explicitely the midpoint algebra $(\boldsymbol{S}, \oplus)$, we fix two distinct points $o, 1 \in \boldsymbol{S}$; then $(\boldsymbol{S}, o, \oplus)$ is a midpoint algebra with zero, and by 6.3.2 the line $F = [o1]$ generates its subalgebra

[4] By 5.2.1 we infer that the reflection with respect to a is a homothety with center a (M.M.).

[5] The term "Th / Mid" indicates midpoint planes, to which the theorem refers.

$(F, o, \oplus)$. The latter yields automatically the group $\mathfrak{F}_+ = (F, o, +)$ with the operator $\frac{1}{2}$ (see § 5.3). We now complete the pair $(o, 1)$ to a system of parameters $(o, 1, e)$, and, on the basis of the axioms for parallelity planes, we construct the multiplicative loop $\mathfrak{F}.(o, 1, e) = (F, o, 1, \cdot)$ with the scalar o as zero. Let us now sketch a proof of distributivity law. Clearly, the axiom 6.3.1 plays the essential role in this proof, since it is the only axiom connecting the operation $\oplus$ with the relation $\|$. By 6.3.1, we get at once

Th 6.3.3. $\sim L(abc) \wedge a \oplus d = b \oplus c \Rightarrow P(abcd)$,
Mid

whereafter

Th 6.3.4. $P(abcd) \Rightarrow \sim L(abc) \wedge a \oplus d = b \oplus c$.
Mid

Proof. $P(abcd)$ implies $\sim L(abc)$; we have also $a \oplus d' = b \oplus c$ for some point d'. Thus, by 6.3.3, we get $P(abcd')$. Consequently $d = d'$, which completes the proof. ∎

We can now extend the relation of parallelogram, P, to the following relation P_M:

Def 6.3.5. $P_M(abcd) \Leftrightarrow a \oplus d = b \oplus c$.[6]

The statements 6.3.3 and 6.3.4 yield

Th 6.3.6. $P(abcd) \Leftrightarrow \sim L(abc) \wedge P_M(abcd)$.
Mid

The definition 6.3.5 enables us to express the statement 5.4.5 in the form

Th des$_M$. $P_M(abpq) \wedge P_M(abrs) \Rightarrow P_M(pqrs)$.

In view of 6.3.6, the set of parallelograms for the parallelity plane $(S, \|)$ coincides with the set of *plane* parallelograms for the midpoint algebra $(S, \oplus)$; however, we now add *linear* parallelograms. This justifies the term "des$_M$" used for the last statement (compare theorem des″ in § 3.2). Of course, theorem des$_M$ is stronger than the former axiom des″, and therefore it yields the invariance of midpoint under

[6] Evidently, this is an alternative way of writing the formula $\overrightarrow{ab} \equiv \overleftarrow{cd}$.

some parallel projections:

Th 6.3.7. $K \neq K' \wedge a \in K, K' \wedge a \neq b, c \in K \wedge a \neq b', c' \in K' \wedge bb' \| cc' \wedge b = a \oplus c \Rightarrow b' = a \oplus c'$.

P r o o f. Assume the antecedent of the implication and let $d = b \ominus b'$ (see fig. 49). Then

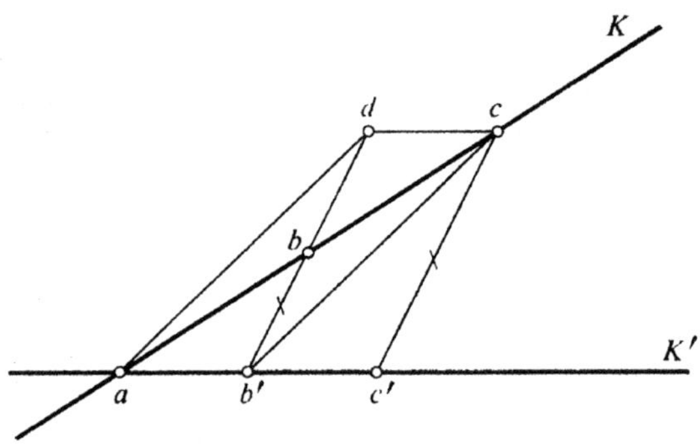

Fig. 49

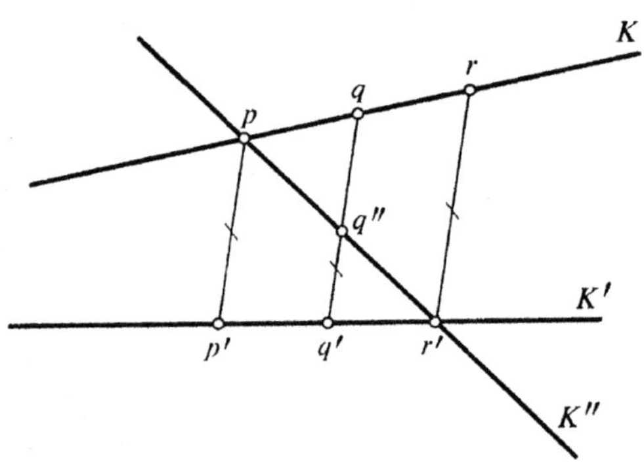

Fig. 50

(1) $P_M(ab'dc)$,

which, together with $\sim L(ab'd)$, gives in turn $P(ab'dc)$ and

(2) $P(dcb'c')$.

From (1) and (2) it follows that $P_M(ab'b'c')$, whence $b' = a \oplus c'$. ∎

As a direct consequence of 6.3.7 (see fig. 50), we get invariance of midpoint under parallel projections of an arbitrary line K onto an arbitrary line K'. In particular, it concerns two arbitrary distinct lines

$K, K' \ni o$, and results in the left distributivity law

$$u \cdot (x \oplus y) = u \cdot x \oplus u \cdot y \qquad \text{(see § 4.3)},$$

which is clearly equivalent to the law

$$u \cdot (x+y) = u \cdot x + u \cdot y.$$

Thus we already have the left quasi-field $\mathfrak{F}(o, 1, e) = (F, o, 1, +, \cdot)$. Evidently $\mathfrak{F}_+ \in \mathbf{G}_{\left(\frac{1}{2}\right)}$.

Applying des_M, we get at once an equation of any line L non-parallel to F. Indeed, by the notation of fig. 51,

$$\boldsymbol{P}(ovpq) \text{ and } \boldsymbol{P}\big(pq(u \cdot x)y\big),$$

whence $\boldsymbol{P}_M\big(ov(u \cdot x)y\big)$, i.e. $o \oplus y = u \cdot x \oplus v$; thus $y = u \cdot x + v$.

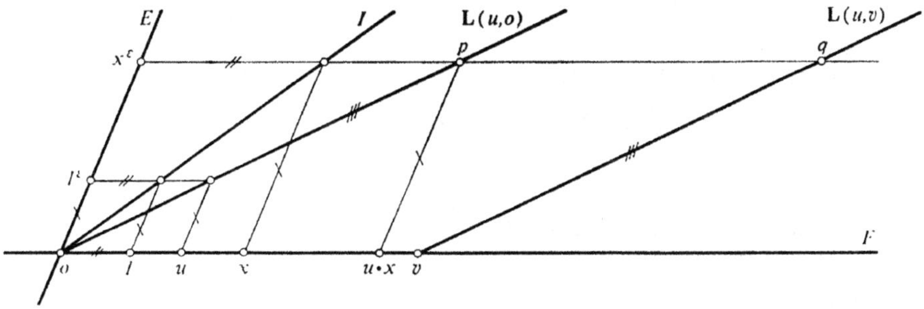

Fig. 51

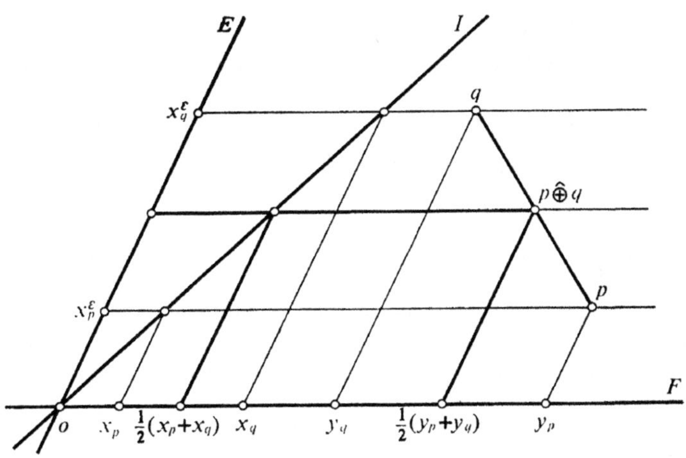

Fig. 52

By means of the equation of a line we obtain the laws 4.4.5(ii) and 0.2.2str. The invariance of midpoint under parallel projections enables us to describe the operation $\oplus$ in terms of $+$:

Th 6.3.8. $\quad \Pi(p \oplus q) = \frac{1}{2}\,(\Pi(p) + \Pi(q))$
Mid

(see fig. 52). This completes the process of coordinatization.

The class of all the midpoint planes will be denoted by the symbol Mid$\mathscr{A}$. The strengthening of the axiom system AX 6.3 by Des or Papp results respectively in the class of Desarguean midpoint planes, Mid Des $\mathscr{A}$, or that of Pappian midpoint planes, Mid Papp $\mathscr{A}$. Evidently

$$\text{Mid } \mathscr{A} \supseteq \text{Mid Des } \mathscr{A} \supseteq \text{Mid Papp } \mathscr{A}.$$

6.4. REPRESENTATION THEOREM[7]

Let us consider the following three classes of quasi-fields:

$$\mathbf{QF}_{\left(\frac{1}{2}\right)} = \{\mathfrak{F} \in \mathbf{QF}\colon \mathfrak{F}_+ \in \mathbf{Q}_{\left(\frac{1}{2}\right)}\},$$
$$\mathbf{F}_{\left(\frac{1}{2}\right)} = \{\mathfrak{F} \in \mathbf{F}\colon \mathfrak{F}_+ \in \mathbf{G}_{\left(\frac{1}{2}\right)}\},$$
$$\text{com}\,\mathbf{F}_{\left(\frac{1}{2}\right)} = \{\mathfrak{F} \in \text{com}\mathbf{F}\colon \mathfrak{F}_+ \in \mathbf{G}_{\left(\frac{1}{2}\right)}\}.$$

Of course

$$\mathbf{QF}_{\left(\frac{1}{2}\right)} \supseteq \mathbf{F}_{\left(\frac{1}{2}\right)} \supseteq \text{com}\,\mathbf{F}_{\left(\frac{1}{2}\right)}.\text{[8]}$$

Let us now connect the classes of planes: Mid $\mathscr{A}$, Mid Des $\mathscr{A}$, and Mid Papp $\mathscr{A}$ with the classes of algebras: $\mathbf{QF}_{\left(\frac{1}{2}\right)}$, $\mathbf{F}_{\left(\frac{1}{2}\right)}$, and com$\,\mathbf{F}_{\left(\frac{1}{2}\right)}$ by the representation theorem. Consider an arbitrary algebra $\mathfrak{F} = (F, 0, 1, +, \cdot) \in \mathbf{QF}_{\left(\frac{1}{2}\right)}$. For any $a, b \in F^2$, let

---

[7] M.M.

[8] The class $\mathbf{F}_{\left(\frac{1}{2}\right)}$ [illegible]ncides with the class of fields of characteristic $\neq 2$. Indeed, the condition $1+1 \neq 0$ is equivalent to the condition $\alpha+\alpha = 0 \Rightarrow \alpha = 0$ (comp. 5.1.2). Making use of extendability of the group $\mathfrak{F}_+$ to a left quasi-field, one can easily prove that $\mathfrak{F}_+ \in \mathbf{G}_{\left(\frac{1}{2}\right)} \Leftrightarrow \text{char } \mathfrak{F} \neq 2$ (compare theorem 6.1.1 together with the remark following it). Obviously the same reasoning is valid for the class $\mathbf{QF}_{\left(\frac{1}{2}\right)}$ (M.M.).

Def 6.4.1. $a \oplus_{\mathfrak{F}} b = \frac{1}{2}(a+b)$.[9]

Evidently, in the Cartesian structure

$$\underline{\mathfrak{A}}(\mathfrak{F}) = (F^2, \oplus_{\mathfrak{F}}, \|_{\mathfrak{F}})$$

all the axioms for midpoint planes are satisfied. Thus we have the following

Theorem 6.4.2 (i) $\mathfrak{F} \in \mathbf{QF}_{\left(\frac{1}{2}\right)} \quad \underline{\mathfrak{A}}(\mathfrak{F}) \in \text{Mid}\,\mathscr{A}$,

(ii) $\mathfrak{F} \in \mathbf{F}_{\left(\frac{1}{2}\right)} \Rightarrow \underline{\mathfrak{A}}(\mathfrak{F}) \in \text{Mid Des}\,\mathscr{A}$,

(iii) $\mathfrak{F} \in \text{com}\,\mathbf{F}_{\left(\frac{1}{2}\right)} \Rightarrow \underline{\mathfrak{A}}(\mathfrak{F}) \in \text{Mid Papp}\,\mathscr{A}$.

P r o o f. (i) is obvious; (ii) follows from (i) combined with 4.0.6(ii); (iii) follows from (i) combined with 4.0.6(iii). ∎

Let now $\mathfrak{A} = (S, \oplus, \|)$ be a midpoint plane. Let $\mathfrak{F} = (F, o, 1, +, \cdot)$ be the quasi-field with a basis $(o, 1, e)$ (see § 6.3) and let $\underline{\mathfrak{A}}(\mathfrak{F}) = (F^2, \oplus_{\mathfrak{F}}, \|_{\mathfrak{F}})$ be the Cartesian midpoint plane over $\mathfrak{F}$. The discussion in § 6.3 leads to the following theorem:

Th 6.4.3. $\mathfrak{A} \underset{\textbf{Mid}}{\simeq} \underline{\mathfrak{A}}(\mathfrak{F})$

and the function Π is an isomorphism.

As a consequence of 6.4.3, we obtain the representation theorem:

Theorem 6.4.4 (i) $\text{Mid}\mathscr{A} \underset{\text{iso}}{\subseteq} \{\underline{\mathfrak{A}}(\mathfrak{F}) : \mathfrak{F} \in \mathbf{QF}_{\left(\frac{1}{2}\right)}\} \subseteq \text{Mid}\,\mathscr{A}$,

(ii) $\text{Mid Des}\,\mathscr{A} \underset{\text{iso}}{\subseteq} \{\underline{\mathfrak{A}}(\mathfrak{F}) : \mathfrak{F} \in \mathbf{F}_{\left(\frac{1}{2}\right)}\} \subseteq \text{Mid Des}\,\mathscr{A}$,

(iii) $\text{Mid Papp}\,\mathscr{A} \underset{\text{iso}}{\subseteq} \{\underline{\mathfrak{A}}(\mathfrak{F}) : \mathfrak{F} \in \text{com}\mathbf{F}_{\left(\frac{1}{2}\right)}\}$

$\subseteq \text{Mid Papp}\,\mathscr{A}$.

P r o o f. (i) follows from 6.4.2(i); (ii) follows from 4.3.9 combined with 6.4.2(ii); (iii) follows from 4.3.10 combined with 6.4.2(iii). ∎

[9] Compare § 0.6.

CHAPTER 7

ALGEBRAIC STRUCTURES WITH A LINEAR ORDERING[1]

7.1. LINEAR ORDERS AND BETWEENNESS RELATIONS

Consider a nonempty set G, a binary relation $\leqslant$ and a ternary relation $\boldsymbol{B}$ on G. The relation $\leqslant$ is said to be a *(linear) order* in G if and only if the structure $(G, \leqslant)$ satisfies the following three axioms:

Ax 7.1.1. $a \leqslant b \vee b \leqslant a$ (*connectivity*),

Ax 7.1.2. $a \leqslant b \wedge b \leqslant a \Rightarrow a = b$ (*antisymmetry*),

Ax 7.1.3. $a \leqslant b \wedge b \leqslant c \Rightarrow a \leqslant c$ (*transitivity*).

The structure $(G, \leqslant)$ is then referred to as a *linearly ordered set*. As an immediate consequence of 7.1.1 we obtain

Th 7.1.4. $a \leqslant a$ (*reflexivity*).

Let

Def 7.1.5. $a \geqslant b \Leftrightarrow b \leqslant a$.

Evidently, if the relation $\leqslant$ is a linear order in G, then its converse, $\geqslant$, is too.

The relation $\boldsymbol{B}$ is said to be a *betweenness relation on* G whenever the following five axioms are satisfied in $(G, \boldsymbol{B})$:

Ax 7.1.6. $\boldsymbol{B}(abc) \vee \boldsymbol{B}(bca) \vee \boldsymbol{B}(cab)$ (*connectivity*),

Ax 7.1.7. $\boldsymbol{B}(aba) \Rightarrow a = b$,

Ax 7.1.8. $\boldsymbol{B}(abc) \Rightarrow \boldsymbol{B}(cba)$ (*symmetry*),

Ax 7.1.9. $\boldsymbol{B}(abc) \wedge \boldsymbol{B}(acd) \Rightarrow \boldsymbol{B}(bcd)$ (*inner transitivity* I),

Ax 7.1.10. $\boldsymbol{B}(abc) \wedge \boldsymbol{B}(bcd) \wedge b \neq c \Rightarrow \boldsymbol{B}(acd)$ (*outer transitivity*).

The following theorems describe further properties of $\boldsymbol{B}$. Their proofs are, in general, based on 7.1.6.

---

[1] See Szmielew [77].

Th 7.1.11. $B(aaa)$.

Proof. $B(aaa) \vee B(aaa) \vee B(aaa) \mapsto B(aaa)$. ■

Th 7.1.12. $B(aab)$ (*reflexivity*).

Proof. By 7.1.7, 7.1.8, and 7.1.11

$$B(aab) \vee B(aba) \vee B(baa) \mapsto B(aab) \vee a = b \mapsto B(aab). ■$$

Th 7.1.13. $B(abc) \wedge B(acb) \Rightarrow b = c$ (*antisymmetry*).

Proof. By 7.1.9 and 7.1.7

$$B(abc) \wedge B(acb) \mapsto B(bcb) \mapsto b \doteq c. ■$$

Th 7.1.14. $B(abd) \wedge B(bcd) \mapsto B(acd)$ (*inner transitivity* II).

Proof. By 7.1.8, 7.1.9, 7.1.13, and 7.1.12

$$B(cda) \wedge B(abd) \wedge B(bcd) \mapsto B(cdb) \wedge B(bcd) \mapsto c = d \mapsto B(acd)$$

and

$$B(dac) \wedge B(abd) \wedge B(bcd) \mapsto B(dac) \wedge B(abc) \wedge B(abd) \wedge B(bcd)$$
$$\mapsto B(dab) \wedge B(abd) \wedge B(bcd) \mapsto a = b \wedge B(bcd) \mapsto B(acd). ■$$

Th 7.1.15. $a \neq b \wedge B(abc) \wedge B(abd) \Rightarrow B(acd) \vee B(adc)$ (*outer smoothness* I).

Proof. By 7.1.9, 7.1.8 and 7.1.13

$$B(dac) \wedge B(abc) \wedge B(abd) \mapsto B(bad) \wedge B(abd) \mapsto a = b. ■$$

Th 7.1.16. $a \neq b \wedge B(abc) \wedge B(abd) \Rightarrow B(bcd) \vee B(bdc)$ (*outer smoothness* II).

Proof. By 7.1.15 and 7.1.9

$$a \neq b \wedge B(abc) \wedge B(abd) \mapsto B(abc) \wedge B(abd) \wedge (B(acd) \vee B(adc))$$
$$\mapsto B(bcd) \vee B(bdc). ■$$

Let us notice that in the proofs of 7.1.11–7.1.16 we have not used the axiom 7.1.10.[2]

We are now going to establish the connections between the linear orders and the betweenness relations on G. In this discussion we assume

---

[2] It will be essential in § 7.2.

the cardinality of G to be at least 3. Let us consider a linear order $\leqslant$ in G and the following relation $\boldsymbol{B}$:

Def 7.1.17. $\boldsymbol{B}(abc) \Leftrightarrow a \leqslant b \leqslant c \vee a \geqslant b \geqslant c.$

By 7.1.1.–7.1.5 it is easy to check that $\boldsymbol{B}$ is a betweenness relation on G. Thus the formula 7.1.17 determines a functor Φ,

$$\Phi(\leqslant) = \boldsymbol{B},$$

which assigns to every linear order in G some betweenness relation on G.[3] Of course

$$\Phi(\leqslant) = \Phi(\geqslant).$$

Let

$$\bar{\Phi}(\{\leqslant, \geqslant\}) = \Phi(\leqslant) = \Phi(\geqslant).$$

The function $\bar{\Phi}$ has unordered pairs of mutually converse linear orders as arguments; as will be shown, $\bar{\Phi}$ is already a bijection.

Consider now a betweenness relation $\boldsymbol{B}$ on G and let $a, b, c \in G$. Assume

(i) $\neq (abc) \wedge \boldsymbol{B}(abc)$

(see fig. 53). In turn, consider the relation $\leqslant$ defined as follows:

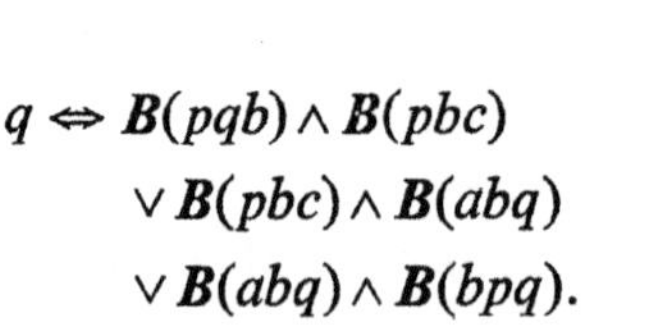

Fig. 53

Def 7.1.18.
$$\begin{aligned} p \leqslant q \Leftrightarrow\ & \boldsymbol{B}(pqb) \wedge \boldsymbol{B}(pbc) \\ & \vee \boldsymbol{B}(pbc) \wedge \boldsymbol{B}(abq) \\ & \vee \boldsymbol{B}(abq) \wedge \boldsymbol{B}(bpq). \end{aligned}$$

[3] In fact, Φ is a functor which assigns to every linearly ordered set $(G, \leqslant)$ a structure $(G, \boldsymbol{B})$ satisfying 7.1.6–7.1.10. However, since the universe G is fixed, it is more convenient to deal with the relations $\leqslant$ and $\boldsymbol{B}$ themselves instead of the corresponding structures.

We are going to show that $\leqslant$ is a linear order in G. Let

(ii) $A = \{p : \boldsymbol{B}(pbc)\}, \quad C = \{p : \boldsymbol{B}(abp)\},$

$$p \underset{A}{\leqslant} q \Leftrightarrow p, q \in A \wedge \boldsymbol{B}(pqb), \quad p \underset{C}{\leqslant} q \Leftrightarrow p, q \in C \wedge \boldsymbol{B}(pbq).$$

Then

(iii) $p \leqslant q \Leftrightarrow p \underset{A}{\leqslant} q \vee p \in A \wedge q \in C \vee p \underset{C}{\leqslant} q.$

Basing on 7.1.6–7.1.16 (in fact we do not use 7.1.7, 7.1.11, and 7.1.15) and (i), we easily show that

(iv) $A \cup C = G$ and $A \cap C = \{b\}$,

(v) the relation $\underset{A}{\leqslant}$ is a linear order in A and $p \underset{A}{\leqslant} b$ for every $p \in A$,

(vi) the relation $\underset{C}{\leqslant}$ is a linear order in C and $b \underset{C}{\leqslant} p$ for every $p \in C$.

From the conditions (iii)–(vi) it follows easily that the relation $\leqslant$ is a linear order in G. Thus the formula 7.1.18 determines a functor Ψ,

$$\Psi(\boldsymbol{B}; a, b, c) = \leqslant,$$

which assigns to every betweenness relation $\boldsymbol{B}$ on G and every triple (a, b, c) satisfying (i), a certain linear order in G.[4] Obviously,

$$\text{if } \Psi(\boldsymbol{B}; a, b, c) = \leqslant, \text{ then } \Psi(\boldsymbol{B}; c, b, a) = \geqslant.$$

Let

$$\overline{\Psi}(\boldsymbol{B}; a, b, c) = \{\leqslant, \geqslant\}.$$

The function $\overline{\Psi}$ has unordered pairs of mutually converse linear orders as values; this function will turn out to be independent of its point arguments.

Let us establish the connection between $\overline{\Phi}$ and $\overline{\Psi}$:

Theorem 7.1.19. *Let* $\{\leqslant, \geqslant\}$ *be the pair of mutually converse linear orders in* G *and let* $\boldsymbol{B}$ *be a betweenness relation on* G.

Then

(*) $\overline{\Phi}(\{\leqslant, \geqslant\}) = \boldsymbol{B} \Leftrightarrow \overline{\Psi}(\boldsymbol{B}; a, b, c) = \{\leqslant, \geqslant\}$

for arbitrary $a, b, c \in G$ *satisfying the condition* (i).

P r o o f. $\Rightarrow$ Assume (i) and

(1) $\overline{\Phi}(\{\leqslant, \geqslant\}) = \boldsymbol{B}.$

---

[4] See footnote 3.

Let

(2) $\Psi(\boldsymbol{B}; a, b, c) = \leqslant'.$

Then $\leqslant'$ is a linear order. Let us show that, under the additional assumption

(3) $a \leqslant b,$

which implies (by (i), (1), and 7.1.17)

(4) $b \leqslant c,$

the order $\leqslant'$ coincides with $\leqslant$. Take arbitrary $p, q \in G$. By the formulae (1)–(4) together with the axioms 7.1.1–7.1.3 and definitions 7.1.17, 7.1.18,

$$p \leqslant q \leqslant b \mathrel{|\!\!\Rightarrow} p \leqslant q \leqslant b \leqslant c \mathrel{|\!\!\Rightarrow} \boldsymbol{B}(pbc) \wedge \boldsymbol{B}(qbc) \wedge \boldsymbol{B}(pqb)$$
$$\mathrel{|\!\!\Rightarrow} p \leqslant' q,$$
$$p \leqslant b \leqslant q \mathrel{|\!\!\Rightarrow} p \leqslant b \leqslant c \wedge a \leqslant b \leqslant q \mathrel{|\!\!\Rightarrow} \boldsymbol{B}(pbc) \wedge \boldsymbol{B}(abq)$$
$$\mathrel{|\!\!\Rightarrow} p \leqslant' q,$$
$$b \leqslant p \leqslant q \mathrel{|\!\!\Rightarrow} a \leqslant b \leqslant p \leqslant q \mathrel{|\!\!\Rightarrow} \boldsymbol{B}(abp) \wedge \boldsymbol{B}(abq) \wedge \boldsymbol{B}(bpq)$$
$$\mathrel{|\!\!\Rightarrow} p \leqslant' q.$$

Since under the assumption $p \leqslant q$ the above three cases are the only possible, hence in any case

$$p \leqslant q \Rightarrow p \leqslant' q,$$

and therefore

$$p \leqslant q \Leftrightarrow p \leqslant' q.$$

Thus $\leqslant' = \leqslant$ and $\Psi(\boldsymbol{B}; a, b, c) = \leqslant$. Replacing the additional assumption (3) by $a \geqslant b$, we obtain $\Psi(\boldsymbol{B}; a, b, c) = \geqslant$. In both cases $\overline{\Psi}(\boldsymbol{B}; a, b, c) = \{\leqslant, \geqslant\}$.

$\Leftarrow$ Assume (i) and

(5) $\Psi(\boldsymbol{B}; a, b, c) = \leqslant.$

Let

(6) $\overline{\Phi}(\{\leqslant, \geqslant\}) = \boldsymbol{B}'.$

Then $\boldsymbol{B}'$ is a betweenness relation. Let us show that $\boldsymbol{B}' = \boldsymbol{B}$. In this proof we shall use the following abbreviation:

$$\boldsymbol{B}(p_1 \dots p_n) \Leftrightarrow \boldsymbol{B}(p_i p_j p_k) \quad \text{for } i \leqslant j \leqslant k \ (i, j, k = 1, \dots, n).$$

Take arbitrary $p, q, r \in G$. By the formulae (ii), (5), and (6) together with the axioms and theorems 7.1.5–7.1.16 and definitions 7.1.18 and 7.1.17,

$$\begin{aligned}&\boldsymbol{B}(pqr) \wedge p, r \in A \mapsto \boldsymbol{B}(pqrbc) \vee \boldsymbol{B}(rqpbc)\\ &\mapsto p \leqslant q \leqslant r \vee p \geqslant q \geqslant r \mapsto \boldsymbol{B}'(pqr),\end{aligned}$$

and similarly

$$\begin{aligned}&\boldsymbol{B}(pqr) \wedge p, r \in C \mapsto \boldsymbol{B}'(pqr);\\ &\boldsymbol{B}(pqr) \wedge p, q \in A \wedge r \in C \mapsto \boldsymbol{B}(pqbc) \wedge \boldsymbol{B}(abr)\\ &\mapsto p \leqslant q \leqslant b \leqslant r \mapsto p \leqslant q \leqslant r \mapsto \boldsymbol{B}'(pqr),\end{aligned}$$

and similarly

$$\begin{aligned}&\boldsymbol{B}(pqr) \wedge p, q \in C \wedge r \in A \mapsto \boldsymbol{B}'(pqr),\\ &\boldsymbol{B}(pqr) \wedge q, r \in A \wedge p \in C \mapsto \boldsymbol{B}'(pqr),\\ &\boldsymbol{B}(pqr) \wedge q, r \in C \wedge p \in A \mapsto \boldsymbol{B}'(pqr).\end{aligned}$$

Since. by (iv), there are no other possibilities, we infer that

$$\boldsymbol{B}(pqr) \Rightarrow \boldsymbol{B}'(pqr),$$

which implies

$$\boldsymbol{B}(pqr) \Leftrightarrow \boldsymbol{B}'(pqr).$$

Hence $\boldsymbol{B}' = B$. Assuming $\Phi(\boldsymbol{B}; a, b, c) = \geqslant$, instead of (5), we come to the same conclusion. Thus any case $B' = B$ and $\overline{\Phi}(\{\leqslant, \geqslant\}) = B$. ■

Let us now consider an arbitrary betweenness relation $\boldsymbol{B}$ and elements $a_i, b_i, c_i \in G$, $i = 1, 2$, satisfying the conditions

$$\neq (a_i b_i c_i) \wedge \boldsymbol{B}(a_i b_i c_i) \quad \text{for } i = 1, 2.$$

By theorem 7.1.19

$$\begin{aligned}&\overline{\Psi}(\boldsymbol{B}; a_1, b_1, c_1) = \{\leqslant, \geqslant\} \mapsto \overline{\Phi}(\{\leqslant, \geqslant\}) = \boldsymbol{B}\\ &\mapsto \overline{\Psi}(\boldsymbol{B}; a_2, b_2, c_2) = \{\leqslant, \geqslant\}.\end{aligned}$$

Thus

$$\overline{\Psi}(\boldsymbol{B}; a_1, b_1, c_1) = \overline{\Psi}(\boldsymbol{B}; a_2, b_2, c_2) = \overline{\Psi}(\boldsymbol{B}),$$

and the equivalence $(*)$ in theorem 7.1.19 can be reformulated as follows:

Th 7.1.20. $\overline{\Phi}(\{\leqslant, \geqslant\}) = \boldsymbol{B} \Leftrightarrow \overline{\Psi}(\boldsymbol{B}) = \{\leqslant, \geqslant\}.$

By 7.1.20, the functors $\bar{\Phi}$ and $\bar{\Psi}$ are mutually inverse bijections from the class of all the pairs of mutually converse linear orders in G to the class of all the betweenness relations on G, and conversely.

If $\bar{\Phi}(\{\leqslant, \geqslant)\} = \boldsymbol{B}$ (or, equivalently, $\bar{\Psi}(\boldsymbol{B}) = \{\leqslant, \geqslant\}$), then the pair of linear orders $\{\leqslant, \geqslant\}$ and the betweenness relation $\boldsymbol{B}$ are said to be *adjoint* each other. Such two objects $\{\leqslant, \geqslant\}$ and $\boldsymbol{B}$ are mutually definable, because, on the one hand, 7.1.17 determines $\boldsymbol{B}$ by means of $\{\leqslant, \geqslant\}$ and, on the other hand, by 7.1.18, the pair $\{\leqslant, \geqslant\}$ can be defined by means of $\boldsymbol{B}$ and inessential parameters a, b, c. Hence, if the pair of linear orders $\{\leqslant, \geqslant\}$ is adjoint to a betweenness relation $\boldsymbol{B}$, then the structure $(G, \{\leqslant, \geqslant\}, \boldsymbol{B})$ is a common definitional expansion of the structures $(G, \{\leqslant, \geqslant\})$ and $(G, \boldsymbol{B})$, and consequently these two structures are definitionally equivalent. This completely describes the connection between each of two ordered sets $(G, \leqslant)$ and $(G, \geqslant)$ and the structure $(G, \boldsymbol{B})$ adjoint to them. Since already the linear order $\leqslant (\geqslant)$ itself generates the betweenness relation $\boldsymbol{B}$, we shall also say that the relation B is adjoint to $\leqslant (\geqslant)$.

Let us now consider a linearly ordered set $(G, \leqslant)$ and the betweenness relation B adjoint to $\leqslant$. A bijection $\varphi\colon G \to G$ is said to be (weakly) increasing[5] whenever

$$a \leqslant b \Rightarrow a^{\varphi} \leqslant b^{\varphi} \qquad \text{for } a, b \in G,$$

i.e. φ preserves the relation $\leqslant$, and so the relation $\geqslant$ as well. A bijection φ is said to be (weakly) decreasing whenever

$$a \leqslant b \Rightarrow a^{\varphi} \geqslant b^{\varphi} \qquad \text{for } a, b \in G,$$

i.e. φ permutes the pair $\{\leqslant, \geqslant\}$. A bijection is called monotone whenever it is either increasing or decreasing. Thus a bijection is monotone if and only if it preserves the pair $\{\leqslant, \geqslant\}$, i.e. preserves the relation $\boldsymbol{B}$. Hence φ is monotone whenever

$$\boldsymbol{B}(abc) \Rightarrow \boldsymbol{B}(a^{\varphi}b^{\varphi}c^{\varphi}) \qquad \text{for} \quad a, b, c \in G.$$

A linear order $\leqslant$ is said to be *continuous*[6] if the following additional axiom is satisfied in $(G, \leqslant)$:

---

[5] We restrict ourselves to bijections.

[6] M.M.—from the author's monographic lecture of 1968/69.

Ax Co. $X_1, X_2 \neq \emptyset \wedge X_1 \cup X_2 = G \wedge X_1 \cap X_2 = \emptyset \wedge X_1 \leqslant X_2$
$\Rightarrow \exists b\ X_1 \leqslant b \leqslant X_2$.[7]

The pair of sets (X_1, X_2) satisfying the antecedent of the above implication is called a *Dedekind section in* $(G, \leqslant)$; the axiom Co is known as the *continuity axiom.* Let us emphasize that this axiom cannot be expressed in the first order language.

A betweenness relation $\boldsymbol{B}$ on G is said to be *continuous* if the following axiom is satisfied in $(G, \boldsymbol{B})$:

Ax Co′. $\exists a\, \boldsymbol{B}(aX_1X_2) \Rightarrow \exists b\, \boldsymbol{B}(X_1 bX_2)$.[8]

Theorem 7.1.21.[9] *Let* $(G, \leqslant)$ *be a linearly ordered set and let* $\boldsymbol{B}$ *be the betweenness relation adjoint to* $\leqslant$. *The order* $\leqslant$ *is continuous if and only if the relation* $\boldsymbol{B}$ *is continuous*; *i.e. on the basis of* 7.1.1–7.1.3 *and* 7.1.6–7.1.10

$$\mathrm{Co} \mathrel{\substack{\models\!\Rightarrow \\ \Leftarrow\!\models}} \mathrm{Co}'.$$

P r o o f. Co $\models\!\Rightarrow$ Co′. If $X_1 = \emptyset$ or $X_2 = \emptyset$, then $\boldsymbol{B}(X_1 bX_2)$ for every $b \in G$.

If $X_1 = \{p_1\}$ $(X_2 = \{p_2\})$, then $\boldsymbol{B}(X_1 bX_2)$ for $b = p_1$ (for $b = p_2$).

So, assume the cardinality of both X_1 and X_2 to be at least 2:

(1) $|X_i| \geqslant 2$ for $i = 1, 2$.

Assume Co and $\boldsymbol{B}(aX_1X_2)$. Then, by 7.1.1–7.1.3 together with 7.1.17,

$$a \leqslant X_1 \leqslant X_2 \vee a \geqslant X_1 \geqslant X_2.$$

Let

(2) $a \leqslant X_1 \leqslant X_2$

(in the case $a \geqslant X_1 \geqslant X_2$ the argumentation is analogous).

Then, by 7.1.2,

$$p, q \in X_1 \cap X_2 \models\!\Rightarrow a \leqslant p \leqslant q \wedge a \leqslant q \leqslant p \models\!\Rightarrow p = q,$$

whence

(3) $X_1 \cap X_2 = \emptyset \vee \exists b\, X_1 \cap X_2 = \{b\}$.

---

[7] The formula $X \leqslant Y$ is an abbreviation of the formula: $\forall x \in X \forall y \in Y\ x \leqslant y$. We shall use similar abbreviations also for the relation $\boldsymbol{B}$.

[8] This form of the continuity axiom is due to Tarski (comp. Tarski [59]).

[9] M.M.—to the end of § 7.1.

If $X_1 \cap X_2 = \{b\}$, then by 7.1.4

$$X_1 \leqslant b \leqslant X_2,$$

and thus

(4) $B(X_1 b X_2)$.

Let

(5) $X_1 \cap X_2 = \emptyset$.

We define $Y_1, Y_2 \subseteq G$:

$$Y_1 = \{q \in G: \exists p \in X_1 \; q \leqslant p\}, \qquad Y_2 = G - Y_1.$$

Of course

(6) $Y_1 \cup Y_2 = G$ and $Y_1 \cap Y_2 = \emptyset$,

whence, by (1) and (5), we obtain at once

(7) $X_1 \subseteq Y_1 \wedge X_2 \subseteq Y_2$,

and so

(8) $Y_1, Y_2 \neq \emptyset$.

We easily check

(9) $Y_1 \leqslant Y_2$.

By (6), (8), (9), and the axiom Co, there is a point $b \in G$ satisfying the condition

$$Y_1 \leqslant b \leqslant Y_2,$$

and thus, by (7) also the condition

$$X_1 \leqslant b \leqslant X_2,$$

which is equivalent to (4). Hence Co $\Rightarrow$ Co′.

Co′ $\Rightarrow$ Co. Assume Co′; let the pair (Y_1, Y_2) be a Dedekind section in $(G, \leqslant)$, i.e. let it satisfy (6), (8), and (9). Fix a point $a \in Y_1$ and define two sets X_1 and X_2:

$$X_1 = \{p \in G: a \leqslant p \leqslant Y_2\}, \qquad X_2 = Y_2.$$

Then $a \leqslant X_1 \leqslant X_2$, and thus

$$B(aX_1X_2),$$

whence, by Co′, there is a point $b \in G$ satisfying the condition

$$\boldsymbol{B}(X_1 b X_2),$$

which is equivalent to the condition

$$X_1 \leqslant b \leqslant X_2 .$$

Applying 7.1.3, we obtain

$$Y_1 \leqslant b \leqslant Y_2 .$$

Thus Co′ $\Rightarrow$ Co. ■

If a nonempty subset of a continuously ordered set has an upper bound, then it has the least upper bound:

Theorem 7.1.22. *Let $\emptyset \neq X \subseteq G$ and let $\leqslant$ be a continuous order in G. Then*

$$X \leqslant a \Rightarrow \exists b_0[X \leqslant b_0 \wedge \forall y(X \leqslant y \leqslant b_0 \Rightarrow y = b_0)].$$

Proof. Let

(1) $\quad X \leqslant a.$

If $a \in X$, we set $b_0 = a$. If

(2) $\quad a \in G - X,$

we define the pair (X_1, X_2):

$$X_1 = \{p \in G : \exists x \in X \; p \leqslant x\}, \qquad X_2 = G - X_1 .$$

Then

(3) $\quad X \subseteq X_1 .$

Using (3), (2), (1), and 7.1.1–7.1.3, we easily prove that the pair (X_1, X_2) is a Dedekind section in $(G, \leqslant)$. Thus, by Co, there is a point b satisfying the condition

(4) $\quad X_1 \leqslant b \leqslant X_2 .$

Moreover, by 7.1.1 and 7.1.2, there exist at most two such points:

$$X_1 \leqslant b_i \leqslant X_2 \quad \text{for } i = 1, 2.$$

By 7.1.1

$$b_1 \leqslant b_2 \vee b_2 \leqslant b_1 .$$

Then the point $b_0 = \min(b_1, b_2)$ satisfies the required conditions. ■

Let us add to the axiom system for continuously ordered sets,

$$\{7.1.1\text{–}7.1.3,\ \mathrm{Co}\},$$

the following axiom, which postulates the *density of order*:

Ax Dens. $p < q \Rightarrow \exists x\ (p < x < q)$.[10]

Then we obtain the following strengthening of the continuity axiom:

Ax Co.$_*$ $X_1, X_2 \neq \emptyset \wedge X_1 \cup X_2 = G \wedge X_1 \cap X_2 = \emptyset \wedge X_1 \leqslant X_2$
$\Rightarrow \dot{\exists} b\ X_1 \leqslant b \leqslant X_2$.

In conclusion, on the basis of the axiom system for dense orders,

$$\{7.1.1\text{–}7.1.3,\ \mathrm{Dens}\},$$

the axioms Co and Co$_*$ are equivalent:

Theorem 7.1.23. *If the order* $\leqslant$ *in* G *is dense, then*

$$\mathrm{Co} \overset{\vdash\Rightarrow}{\underset{\Leftarrow\dashv}{}} \mathrm{Co}_*.$$

7.2. (APPENDIX TO § 7.1) THE BETWEENNESS RELATION ON A SET OF CARDINALITY $\neq 4$

Let us rewrite the axioms for betweenness relation:

Ax 7.1.6. $B(abc) \vee B(bca) \vee B(cab)$,

Ax 7.1.7. $B(aba) \Rightarrow a = b$,

Ax 7.1.8. $B(abc) \Rightarrow B(cba)$,

Ax 7.1.9. $B(abc) \wedge B(acd) \Rightarrow B(bcd)$,

Ax 7.1.10. $B(abc) \wedge B(bcd) \wedge b \neq c \Rightarrow B(acd)$.

Each of these axioms is independent of the others. The independence models are very simple.

Every structure $(G, \emptyset \times \emptyset \times \emptyset)$ of cardinality $|G| \geqslant 1$ is an independence model for the axiom 7.1.6, and every structure $(G, G \times G \times G)$ with $|G| \geqslant 2$ is an independence model for 7.1.7.

Let us now take an arbitrary linearly ordered set $(G, \leqslant)$ with $|G| \geqslant 2$,

[10] Of course, the symbol "$<$" has here the usual meaning:
$a < b \Leftrightarrow a \leqslant b \wedge a \neq b$.

and let

$$B_8(abc) \Leftrightarrow a \leqslant b \leqslant c \vee a > b > c;$$

then the structure (G, B_8) is an independence model for 7.1.8.

Take an arbitrary set

$$G = \{a, b, c, d\}$$

consisting of four distinct elements a, b, c, d. Let

$$B_0 = \{(ppq): p, q \in G\} \cup \{(pqq): p, q \in G\},$$

and let

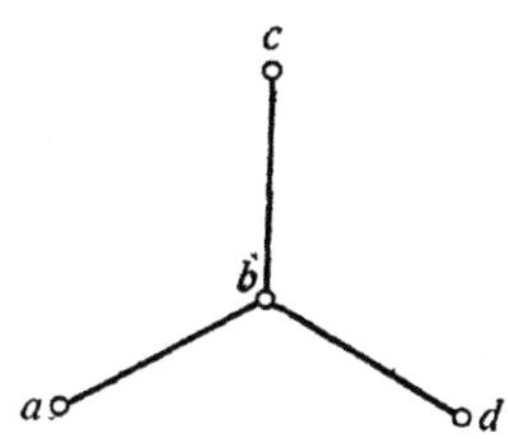

$$B_9 = \{(abc), (cbd), (acd), (abd), (cba), (dbc), (dca). (dba)\}$$

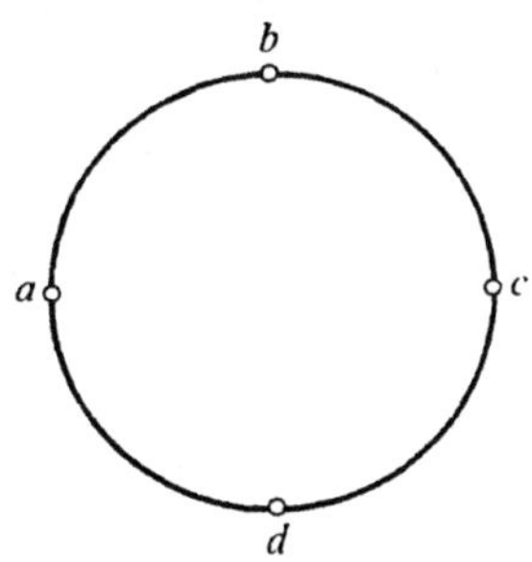

and

$$B_{10} = \{(abc), (bcd), (cda), (dab), (cba), (dcb), (adc), (bad)\}.$$

It is easy to check that the structure

$$\mathfrak{G}_9 = (G, B_0 \cup B_9)$$

is an independence model for 7.1.9, and the structure

$$\mathfrak{G}_{10} = (G, B_0 \cup B_{10})$$

is an independence model for 7.1.10.

The model $\mathfrak{G}_9$ can be easily extended to an independence model (for 7.1.9) of any finite cardinality > 4 or of infinite cardinality. Indeed, let us settle a linear order in G, e.g. let

$$a \leqslant b \leqslant c \leqslant d;$$

then $(G, \leqslant)$ is a linearly ordered set. Take now any linearly ordered set $(F, \leqslant)$ with $F \wedge G = \emptyset$. Setting

$$F \leqslant G,$$

we obtain a linearly ordered set $(F \cup G, \leqslant)$. Consider the structure $(F \cup G, \boldsymbol{B})$ adjoint to $(F \cup G, \leqslant)$ and let

$$\boldsymbol{B}' = \boldsymbol{B} - \{(bcd), (dcb)\} \cup \{(cdb), (dbc)\}.$$

Evidently

$$\boldsymbol{B}' | G \times G \times G = \boldsymbol{B}_0 \cup \boldsymbol{B}_9.$$

The structure $(F \cup G, \boldsymbol{B}')$ is an independence model for the axiom 7.1.9.

We are going to show that, in contrast with 7.1.6–7.1.9, the axiom 7.1.10 has no independence model of cardinality different from 4. To this end we shall prove (basing only on 7.1.6–7.1.9) several transitivity laws for five elements. We start with the inner transitivity law (7.2.1), and finish in the outer transitivity law (7.2.4). Let us recall that the statements 7.1.11–7.1.16 have been proved without the use of 7.1.10 and therefore we can use them now[11]. We shall also use the abbreviation $\boldsymbol{B}(p_1 \ldots p_n)$ introduced in § 7.1.

From 7.1.9 and 7.1.14 we derive directly

Th 7.2.0. $\boldsymbol{B}(abc) \wedge \boldsymbol{B}(acd) \Rightarrow \boldsymbol{B}(abcd)$.

abc
a cd
abcd

By 7.2.0 we get

Th. 7.2.1. $\boldsymbol{B}(abc) \wedge \boldsymbol{B}(acd) \wedge \boldsymbol{B}(ade) \Rightarrow \boldsymbol{B}(abcde)$.

abc
a cd
a de
abcde

[11] Compare footnote 2.

P r o o f. $\quad \boldsymbol{B}(abc) \wedge \boldsymbol{B}(acd) \mathrel{|}\!\Rightarrow \boldsymbol{B}(abcd),$

$$\boldsymbol{B}(acd) \wedge \boldsymbol{B}(ade) \mathrel{|}\!\Rightarrow \boldsymbol{B}(acde),$$
$$\boldsymbol{B}(abd) \wedge \boldsymbol{B}(ade) \mathrel{|}\!\Rightarrow \boldsymbol{B}(abde),$$
$$\boldsymbol{B}(bcd) \wedge \boldsymbol{B}(bde) \mathrel{|}\!\Rightarrow \boldsymbol{B}(bcde),$$

and thus $\boldsymbol{B}(abcde)$. ■

Th 7.2.2. $\quad \boldsymbol{B}(abc) \wedge \boldsymbol{B}(acd) \wedge \boldsymbol{B}(bde) \wedge b \neq c \neq d \Rightarrow \boldsymbol{B}(abcde).$

$$\begin{array}{|l|} abc \\ a\ cd \\ \ \ b\ de \\ \hline abcde \end{array}$$

P r o o f. By 7.2.1 it suffices to show that the assumption implies $\boldsymbol{B}(ade)$. By 7.2.0

$$\boldsymbol{B}(abc) \wedge \boldsymbol{B}(acd) \mathrel{|}\!\Rightarrow \boldsymbol{B}(abcd),$$
$$\boldsymbol{B}(bcd) \wedge \boldsymbol{B}(bde) \mathrel{|}\!\Rightarrow \boldsymbol{B}(bcde).$$

Notice that $\sim \boldsymbol{B}(cae)$ and $\sim \boldsymbol{B}(aec)$; indeed

$$\boldsymbol{B}(cba) \wedge \boldsymbol{B}(cae) \mathrel{|}\!\Rightarrow \boldsymbol{B}(cbae),$$

which, together with $\boldsymbol{B}(bce)$, implies $b = c$; similarly,

$$\boldsymbol{B}(aec) \wedge \boldsymbol{B}(acd) \mathrel{|}\!\Rightarrow \boldsymbol{B}(aecd),$$

which, together with $\boldsymbol{B}(cde)$, implies $c = d$ (see 7.1.8 and 7.1.13). Thus $\boldsymbol{B}(eca)$ and

$$\boldsymbol{B}(edc) \wedge \boldsymbol{B}(eca) \mathrel{|}\!\Rightarrow \boldsymbol{B}(edca) \mathrel{|}\!\Rightarrow \boldsymbol{B}(ade). \ ■$$

Th 7.2.3. $\quad \boldsymbol{B}(abc) \wedge \boldsymbol{B}(acd) \wedge \boldsymbol{B}(cde) \wedge a \neq b \neq c \neq d$
$\Rightarrow \boldsymbol{B}(abcde).$

$$\begin{array}{|l|} abc \\ a\ cd \\ \ \ \ \ cde \\ \hline abcde \end{array}$$

P r o o f. In view of 7.2.2 it suffices to show that the assumption implies $\boldsymbol{B}(bde)$. By 7.2.0

$$\boldsymbol{B}(abc) \wedge \boldsymbol{B}(acd) \mathrel{|}\!\Rightarrow \boldsymbol{B}(abcd).$$

Notice that $\sim \boldsymbol{B}(aec)$ and $\sim \boldsymbol{B}(cae)$; indeed,

$$\boldsymbol{B}(aec) \wedge \boldsymbol{B}(acd) \mathrel{|}\!\Rightarrow \boldsymbol{B}(aecd)$$

and, by 7.2.2, since $a \neq b \neq c$,

$$\boldsymbol{B}(dcb) \wedge \boldsymbol{B}(dba) \wedge \boldsymbol{B}(cae) \mid\!\Rightarrow \boldsymbol{B}(dcbae);$$

since $\boldsymbol{B}(cde)$, in both cases we get $c = d$. Thus $\boldsymbol{B}(ace)$ and

$$\boldsymbol{B}(abc) \wedge \boldsymbol{B}(ace) \mid\!\Rightarrow \boldsymbol{B}(abce),$$

$$\boldsymbol{B}(edc) \wedge \boldsymbol{B}(ecb) \mid\!\Rightarrow \boldsymbol{B}(edcb) \mid\!\Rightarrow \boldsymbol{B}(bde). \blacksquare$$

Th 7.2.4. $\boldsymbol{B}(abc) \wedge \boldsymbol{B}(bcd) \wedge \boldsymbol{B}(cde)$
$\wedge\ a \neq b \neq c \neq d \neq e \neq a \Rightarrow \boldsymbol{B}(abcde).$

abc
bcd
cde
abcde

Proof. In view of 7.2.3, it suffices to show that the assumption implies $\boldsymbol{B}(acd)$. Notice that $\sim \boldsymbol{B}(cea)$; indeed, by 7.2.0

$$\boldsymbol{B}(cde) \wedge \boldsymbol{B}(cea) \mid\!\Rightarrow \boldsymbol{B}(cdea),$$

and, by 7.2.3, since $c \neq d \neq e \neq a$,

$$\boldsymbol{B}(aed) \wedge \boldsymbol{B}(adc) \wedge \boldsymbol{B}(dcb) \mid\!\Rightarrow \boldsymbol{B}(aedcb),$$

which, together with $\boldsymbol{B}(abc)$, implies $b = c$. On the other hand, $\sim \boldsymbol{B}(cae)$; indeed,

$$\boldsymbol{B}(cba) \wedge \boldsymbol{B}(cae) \mid\!\Rightarrow \boldsymbol{B}(cbae),$$

and by 7.2.3, since $e \neq a \neq b \neq c$,

$$\boldsymbol{B}(eab) \wedge \boldsymbol{B}(ebc) \wedge \boldsymbol{B}(bcd) \mid\!\Rightarrow \boldsymbol{B}(eabcd),$$

which, together with $\boldsymbol{B}(cde)$, implies $c = d$. Hence $\boldsymbol{B}(ace)$ and

$$\boldsymbol{B}(abc) \wedge \boldsymbol{B}(ace) \mid\!\Rightarrow \boldsymbol{B}(abce),$$

$$\boldsymbol{B}(edc) \wedge \boldsymbol{B}(eca) \mid\!\Rightarrow \boldsymbol{B}(edca) \mid\!\Rightarrow \boldsymbol{B}(acd). \blacksquare$$

Let us now extend the axiom system 7.1.6–7.1.9 by the following axiom, which excludes universes of cardinality 4.

Ax 7.2.5. $\neq (abcd) \Rightarrow \exists p\ p \neq a, b, c, d.$

On the basis of the strengthened axiom system we can already prove the outer transitivity law for four points (i.e. the axiom 7.1.10):

Th 7.2.6. $\boldsymbol{B}(abc) \wedge \boldsymbol{B}(bcd) \wedge b \neq c \Rightarrow \boldsymbol{B}(acd).$

Proof. The theorem is obvious if any two elements of the set $\{a, b, c, d\}$ are identical. Thus we can assume that $\neq (abcd)$. Then, by the axiom 7.2.5, there exists a $p \neq a, b, c, d$. By 7.1.6 and 7.1.16

(1) $\quad \boldsymbol{B}(pab) \vee \boldsymbol{B}(apb) \vee \boldsymbol{B}(bpc) \vee \boldsymbol{B}(cpd) \vee \boldsymbol{B}(cdp)$.

Let us notice that

$$\boldsymbol{B}(cpb) \wedge \boldsymbol{B}(cba) \Rightarrow \boldsymbol{B}(cpba),$$

whence the third case may be written in the form

$$\boldsymbol{B}(abp) \wedge \boldsymbol{B}(apc).$$

Applying in turn 7.2.4, 7.2.3, 7.2.2, 7.2.3, and 7.2.4, we pass from the disjunction (1) to the following one:

$$\boldsymbol{B}(pabcd) \vee \boldsymbol{B}(apbcd) \vee \boldsymbol{B}(abpcd) \vee \boldsymbol{B}(abcpd) \vee \boldsymbol{B}(abcdp).$$

pab	*apb*	*abp*	*dpc*	*abc*
abc	*a bc*	*a pc*	*d cb*	*bcd*
bcd	*bcd*	*b cd*	*cba*	*cdp*
pabcd	*apbcd*	*abpcd*	*dpcba*	*abcdp*

Hence, in any case $\boldsymbol{B}(abcd)$ and, in particular, $\boldsymbol{B}(acd)$. ∎[12]

As a corollary we obtain

Theorem 7.2.7. *Let G be a nonempty set and let $\boldsymbol{B} \subset G \times G \times G$. If $|G| \neq 4$ and the structure $(G, \boldsymbol{B})$ satisfies the axioms* 7.1.6–7.1.9, *then $\boldsymbol{B}$ is a betweenness relation on G,* i.e. *$(G, \boldsymbol{B})$ satisfies the axiom* 7.1.10 *as well.*

Thus every independence model for the axiom 7.1.10 has cardinality 4. It is easy to check that it is isomorphic to $\mathfrak{G}_{10}$. Hence $\mathfrak{G}_{10}$ is the unique (up to isomorphism) structure which satisfies 7.1.6–7.1.9 but at the same time does not satisfy 7.1.10. The axiom 7.1.10 is needed only to exclude the structure $\mathfrak{G}_{10}$ from the class of ordered sets. On the other hand, though the proof of 7.1.10 on the base of 7.1.6–7.1.9 and 7.2.5 is clear, still it is rather long and strenuous.

7.3. ORDERED GROUPS

A structure $(\mathfrak{G}, \leqslant) = (\mathfrak{G}, 0, +, \leqslant)$[13] consisting of a commutative group $\mathfrak{G} = (G, 0, +)$ and a linear order $\leqslant$ in G is said to be an *ordered*

[12] See Piesyk [77].

[13] We identify the structure $(G, 0, +, \leqslant)$ with the pair $((G, 0, +), \leqslant)$.

group whenever the following axiom is satisfied in $(\mathfrak{G}, \leqslant)$:

Ax 7.3.0. $p \leqslant q \Rightarrow p+a \leqslant q+a$ *(monotony).*

The relation $\leqslant$ is then referred to as a *(linear) order in* $\mathfrak{G}$ (or a *linear ordering of* $\mathfrak{G}$). It is obvious (because of the form of 7.3.0) that the converse relation $\geqslant$ is also an ordering of $\mathfrak{G}$. Hence, with any ordered group we associate the pair of linear orders $\{\leqslant, \geqslant\}$ and thus also the betweenness relation $\boldsymbol{B}$ on G, adjoint to this pair.

The axiom 7.3.0 yields

Th 7.3.1. $p \leqslant q \Rightarrow -p \geqslant -q$.[14]

Proof. Let $a = -(p+q)$. Then

$$p+a = -q \quad \text{and} \quad q+a = -p.$$

Thus

$$p \leqslant q \models p+a \leqslant q+a \models -q \leqslant -p. \blacksquare$$

Consider now the bijections $\sigma^a\colon G \to G$ defined by 0.6.1(i) and the bijection $\mu\colon G \to G$ defined by 0.6.2. Let us prove

Theorem 7.3.2. *Let* $\mathfrak{G} = (G, 0, +)$ *be a commutative group, let* $\leqslant$ *be a linear order in the set* G, *and let* $\boldsymbol{B}$ *be the betweenness relation adjoint to* $\leqslant$. *If* $|G| \geqslant 3$, *then the following conditions are equivalent:*

(o) *The relation* $\leqslant$ *is an ordering of the group* $\mathfrak{G}$.

(i) *All the bijections* σ^a *are increasing.*

(ii) *All the bijections* σ^a *and the bijection* μ *are monotone.*

(iii) *All the bijections* σ^a *and the bijection* μ *preserve the relation* $\boldsymbol{B}$.

Proof. (o) $\rightleftarrows$ (i) is obvious for arbitrary cardinality of G. (i) $\models$ (ii) (for arbitrary cardinality of G): The condition (i) is an alternative form of 7.3.0, while (ii) is an alternative form of 7.3.1. (ii) $\models$ (i): Assume (ii). The bijection σ^0 is always increasing. Let us show that the remaining bijections σ^a are now increasing. Let $a > 0$. Take arbitrary $b \neq 0, a$. Then

$$0 < a < b \quad \text{or} \quad 0 < b < a \quad \text{or} \quad b < 0 < a.$$

[14] Geometrically, the axiom 7.3.0 and the theorem 7.3.1 read respectively as follows: *the relation* $\leqslant$ *is invariant under translations*, and *the reflection with respect to* 0 *permutes the pair* $\{\leqslant, \geqslant\}$.

Let $0 < a < b$. Then either

(1) $\quad -b < (-b)+a < 0,$

when σ^{-b} is increasing, or

(2) $\quad 0 < (-b)+a < -b,$

when σ^{-b} is decreasing. Simultaneously, either

(1′) $\quad -b < -a < 0,$

when μ is decreasing, or

(2′) $\quad 0 < -a < -b,$

when μ is increasing. Since (1′) contradicts (2), and (2′) contradicts (1), there are only two possibilities: either (1) and (1′), and then

$$-b < -a \quad \text{and} \quad (-b)+a < (-a)+a,$$

or (2) and (2′), and then

$$-a < -b \quad \text{and} \quad (-a)+a < (-b)+a.$$

In both cases the bijection σ^a, which was assumed monotone, turns out to be increasing. Assuming $0 < b < a$ or $b < 0 < a$ and reasoning like above, we come to the same conclusion. Thus for every $a > 0$ the bijection σ^a is increasing. For $a < 0$ the proof is analogous.

(ii) $\rightleftarrows$ (iii) is obvious for arbitrary cardinality of G. ∎

In view of 7.3.2, for groups of cardinality $\geqslant 3$, the axiom 7.3.0 is equivalent to the conjunction of the following two sentences:

Ax 7.3.4. $\quad \boldsymbol{B}(pqr) \Rightarrow \boldsymbol{B}(p+a\ q+a\ r+a).$

Ax 7.3.5. $\quad \boldsymbol{B}(pqr) \Rightarrow \boldsymbol{B}(-p\ -q\ -r).$

The structure $(\mathfrak{G}, \boldsymbol{B}) = (G, 0, +, \boldsymbol{B})$ satisfying 7.3.4 and 7.3.5 is definitionally equivalent to $(G, \{\leqslant, \geqslant\})$; it will be also referred to as an *ordered group*. The relation $\boldsymbol{B}$ will be called a *non-oriented order in* $\mathfrak{G}$.

An ordering of a group may be described as well in terms of the so-called positive cone.[15] Consider the structure $(\mathfrak{G}, P) = (G, 0, +, P)$,

[15] M.M.—on the basis of the author's lecture (to the end of § 7.3).

consisting of a commutative group $\mathfrak{G} = (G, 0, +)$ and a set $P \subset G$, satisfying the following axioms:

Ax 7.3.6. $a \in P \vee -a \in P$.

Ax 7.3.7. $a \in P \wedge -a \in P \Rightarrow a = 0$.

Ax 7.3.8. $a \in P \wedge b \in P \Rightarrow a+b \in P$.

By 7.3.6 we obtain

Th 7.3.9. $0 \in P$.

The structure ($\mathfrak{G}$ P) is definitionally equivalent to an ordered group. Indeed, we define the set P in terms of the structure $(\mathfrak{G}, \leqslant)$:

Def 7.3.10. $P = \{x \in G : x \geqslant 0\}$,

and the relation $\leqslant$ in terms of the structure $(\mathfrak{G}, P)$:

Def 7.3.11. $a \leqslant b \Leftrightarrow b-a \in P$.

Then, the following inferences are based on the axioms for commutative groups:

$$\{7.3.0,\ 7.3.10\} \rightleftarrows 7.3.11,$$

and, by means of 7.3.0 and 7.3.10 (or 7.3.11),

$$7.1.1 \rightleftarrows 7.3.6,$$
$$7.1.2 \rightleftarrows 7.3.7,$$
$$7.1.3 \rightleftarrows 7.3.8.$$

The structure $(\mathfrak{G}, P)$ is also called an *ordered group*; the set P is referred to as an *order* or a *positive cone in* $\mathfrak{G}$.

In a commutative group $\mathfrak{G} = (G, 0, +)$ we define as usually the multiplication of an element of the set G by natural numbers:

Def 7.3.12. (i) $0 \cdot a = 0$[16]

(ii) $n \in N - \{0\} \Rightarrow n \cdot a = (n-1) \cdot a + a$.

We have the following

Theorem 7.3.13. *Let* $(G, \leqslant)$ *be an ordered group and let* $a > 0$.

[16] The same symbol is used here to denote the neutral element of $\mathfrak{G}$ and the number $0 \in N$.

Then the set

$$X = \{n \cdot a \colon n \in N\}$$

has no least upper bound.

Proof. Let $X \leqslant b$ for some $b \in G$, i.e.

$$\forall n \in N \; n \cdot a \leqslant b.$$

Then

$$\forall n \in N \; (n+1) \cdot a \leqslant b.$$

Thus, by 7.3.12 and 7.3.0,

$$\forall n \quad n \cdot a \leqslant b-a,$$

i.e. $X \leqslant b-a < b$, whence b is not the least upper bound of X. ∎

An order $\leqslant$ in a group $\mathfrak{G}$ is *Archimedean* whenever the structure $(\mathfrak{G}, \leqslant)$ satisfies the following additional axiom:

Ax Arch. $a > 0 \wedge b > 0 \Rightarrow \exists n \; n \cdot a \geqslant b.$

Let us prove

Theorem 7.3.14. *Let* $(\mathfrak{G}, \leqslant)$ *be an ordered group. If the order* $\leqslant$ *is continuous, then it is Archimedean, i.e.*

$$\mathrm{Co} \models \mathrm{Arch}.$$

Proof. Suppose $\leqslant$ is non-Archimedean, i.e.

$$\exists a, b > 0 \, \forall n \; n \cdot a < b.$$

Then the set $X = \{n \cdot a \colon n \in N\}$ has an upper bound, and thus, by 7.1.22, it has the least upper bound, contrary to 7.3.13. ∎

7.4. (APPENDIX TO § 7.3) ORDERABILITY OF A COMMUTATIVE GROUP WITHOUT CYCLIC ELEMENTS[17]

Let $\mathfrak{G} = (G, 0, +)$ be a commutative group. The group $\mathfrak{G}$ *has no cyclic elements* (i.e. elements of finite order) if and only if

$$(*) \qquad n \cdot a = 0 \Rightarrow n = 0 \vee a = 0.$$

We have the following

[17] M.M.—according to the notes from the seminar conducted by the author in 1973.

Theorem 7.4.1. *An orderable group has no cyclic elements.*

Proof. Let P be an order in the group $\mathfrak{G}$. Take $n \in N$, $a \in G$, and let

(1) $\quad n \cdot a = 0 \wedge n > 0.$

By 7.3.6

$$a \in P \quad \text{or} \quad -a \in P.$$

By 7.3.12, 7.3.9, 7.3.8, and (1)

$$a \in P \Rightarrow (n-1) \cdot a \in P \Rightarrow -a \in P$$

and

$$-a \in P \Rightarrow (n-1) \cdot (-a) \in P \Rightarrow a \in P.$$

Thus, by 7.3.7, $a = 0$. Therefore the group $\mathfrak{G}$ satisfies $(*)$. ∎

We are going to show that the condition $(*)$ is also sufficient for a group to be orderable. We start with two lemmas concerning partial orders.

A set P is referred to as a *partial order* in the group $\mathfrak{G}$ whenever the structure $(\mathfrak{G}, P)$ satisfies the conditions 7.3.7–7.3.9.

Lemma 7.4.2. *Let* $\mathfrak{G} = (G, 0, +)$ *be a group without cyclic elements and let* P *be a partial order in* $\mathfrak{G}$*. Then*

$$x \neq 0 \Rightarrow (\forall n > 0 \;\, n \cdot x \notin P) \vee (\forall n > 0 \;\, n \cdot (-x) \notin P).$$[18]

Proof. If for some $k, l > 0$

$$k \cdot x \in P \wedge l \cdot (-x) \in P,$$

then by 7.3.8

$$k \cdot l \cdot x \in P \wedge -(k \cdot l \cdot x) \in P,$$

whence, by 7.3.7,

$$k \cdot l \cdot x = 0.$$

Therefore, by $(*)$, $x = 0$. ∎

Lemma 7.4.3. *Let* $\mathfrak{G} = (G, 0, +)$ *be a commutative group without cyclic elements and let* $\mathfrak{P}$ *be the class of all the partial orders in* $\mathfrak{G}$*. Then*

$$\forall P \in \mathfrak{P} \forall x \in G - \{0\} \exists P' \in \mathfrak{P} \, [P \subset P' \wedge (x \in P' \vee -x \in P')].$$

[18] This is clearly a strengthening of the axiom 7.3.7.

P r o o f. Let $P \in \mathfrak{P}$ and let $x \neq 0$. By 7.4.2, either

(1) $\forall n > 0 \;\; n \cdot (-x) \notin P$

or

(1′) $\forall n > 0 \;\; n \cdot x \notin P$.

Assume (1). Let

$$P' = \{n \cdot x + p : n \in N \wedge p \in P\}.$$

Then

$$P \subset P' \quad \text{and} \quad x \in P'.$$

The set P' is a partial order in $\mathfrak{G}$; indeed, the set P' satisfies 7.3.9, since N and P satisfy 7.3.9[19]; it satisfies also 7.3.8, since N and P satisfy 7.3.8. Moreover, by (1),

$$n \cdot x + p = 0 \Rightarrow n \cdot (-x) = p \in P \Rightarrow n = 0 \Rightarrow n = 0 \wedge p = 0;$$

thus

(2) $n \cdot x + p = 0 \Rightarrow n = 0 \wedge p = 0$.

If $y \in P' \wedge -y \in P'$, then

$$y = n_1 \cdot x + p_1 \quad \text{and} \quad -y = n_2 \cdot x + p_2$$

for some $n_1, n_2 \in N$ and $p_1, p_2 \in P$. Thus, by (2),

$$(n_1 \cdot x + p_1) + (n_2 \cdot x + p_2) = 0 \Rightarrow (n_1 + n_2) \cdot x + (p_1 + p_2) = 0$$
$$\Rightarrow n_1 + n_2 = 0 \wedge p_1 + p_2 = 0,$$

and therefore $n_1 = n_2 = 0$ and $p_1 = p_2 = 0$ because both N and P satisfy 7.3.7. Thus $y = 0$, whence P' satisfies 7.3.7 as well.

For the case (1′) the proof is analogous. ■

Using Lemma 7.4.3, let us prove

Theorem 7.4.4. *In a commutative group without cyclic elements every partial order can be extended to a linear order.*

P r o o f. Let $\mathfrak{G} = (G, 0, +)$ be a group without cyclic elements, and let α be the ordinal of the set G.[20] Arrange all the elements of G in an α-termed sequence,

$$(0, x_1, \ldots, x_i, x_{i+1}, \ldots), \qquad i < \alpha.$$

---

[19] The set N is the positive cone in the group of integers.

[20] We use the theorem on well-ordered sets.

By 7.4.3 there is an increasing sequence of partial orders in $\mathfrak{G}$,

$$P_1 \subseteq P_2 \subseteq \ldots \subseteq P_i \subseteq P_{i+1} \subseteq \ldots$$

such that

$$\forall i < \alpha \ (x_i \in P_i \vee -x_i \in P_i).$$

We easily check that the set

$$\bar{P} = \bigcup_{i<\alpha} P_i$$

is an order in the group $\mathfrak{G}$. ■

Since the set $P = \{0\}$ is clearly a partial order in $\mathfrak{G}$, as a direct consequence of 7.4.4 we get

Theorem 7.4.5. *Every commutative group $\mathfrak{G}$ without cyclic elements is orderable, i.e. there is a $P \subseteq G$ such that $(\mathfrak{G}, P)$ is an ordered group.*

7.5. ORDERED QUASI-FIELDS

A structure $(\mathfrak{F}, \leqslant) = (F, 0, 1, +, \cdot, \leqslant)$ consisting of a left quasi-field $\mathfrak{F} = (F, 0, 1, +, \cdot)$ and a linear order $\leqslant$ in F[21] is referred to as an *ordered left quasi-field* if it satisfies the axiom 7.3.0 (i.e. the relation $\leqslant$ is an ordering of the additive group $(F, 0, +)$) and the following two axioms:

Ax 7.5.1. $0 \leqslant a \wedge p \leqslant q \Rightarrow p \cdot a \leqslant q \cdot a$ (right monotony).

Ax 7.5.2. $0 \leqslant a \wedge p \leqslant q \Rightarrow a \cdot p \leqslant a \cdot q$ (left monotony).

(Let us recall that the multiplication in a quasi-field need not to be commutative.) Then the relation $\leqslant$ is said to be an *ordering of the quasi-field* $\mathfrak{F}$.

Let $(\mathfrak{F}, \leqslant)$ be an ordered left quasi-field. Then

Th 7.5.3. $0 < 1$.

Proof. By 7.3.0 combined with any one of the axioms 7.5.1 and 7.5.2

$$1 \leqslant 0 \Rightarrow 0 \leqslant -1 \Rightarrow (-1) \cdot (-1) = 1. \ ■$$

---

[21] We identify the structure $(F, 0, 1, +, \cdot, \leqslant)$ with the pair $((F, 0, 1, +, \cdot), \leqslant)$ (comp. footnote 1) (M.M.).

From 7.5.3 it follows immediately that the converse relation $\geqslant$ is not an ordering of the quasi-field $\mathfrak{F}$. Thus at most one of two mutually converse orders $\leqslant$ and $\geqslant$ in the additive group $(F, 0, +)$ is an order in the quasi-field $\mathfrak{F}$.

Let us prove

Theorem 7.5.4. *Let* $\mathfrak{F} = (F, 0, 1, +, \cdot)$ *be a left quasi-field and let* $\leqslant$ *be a linear order in the set* F. *If the relation* $\leqslant$ *is an ordering of the additive group* $(F, 0, +)$, *then the following three conditions are equivalent:*

(i) $\quad 0 \leqslant a \wedge p \leqslant q \Rightarrow p \cdot a \leqslant q \cdot a \wedge a \cdot p \leqslant a \cdot q,$

(ii) $\quad 0 \leqslant a \wedge p \leqslant q \Rightarrow p \cdot a \leqslant q \cdot a,$

(iii) $\quad a \leqslant b \wedge p \leqslant q \Rightarrow a \cdot q + b \cdot p \leqslant a \cdot p + b \cdot q.$

Proof. (i) $\models\!\Rightarrow$ (ii) is obvious.

(ii) $\models\!\Rightarrow$ (iii). Assume (ii). In view of 7.3.0, by left distributivity of the multiplication with respect to the subtraction (see 0.3.4),

$$a \leqslant b \wedge p \leqslant q \models\!\Rightarrow a \leqslant b \wedge 0 \leqslant q - p \models\!\Rightarrow a \cdot (q-p) \leqslant b \cdot (q-p)$$
$$\models\!\Rightarrow a \cdot q - a \cdot p \leqslant b \cdot q - b \cdot p \models\!\Rightarrow a \cdot q + b \cdot p \leqslant a \cdot p + b \cdot q.$$

(iii) $\models\!\Rightarrow$ (i). As particular cases of (iii) (for $p = 0$ or $a = 0$) we get the formulae

$$a \leqslant b \wedge 0 \leqslant q \Rightarrow a \cdot q \leqslant b \cdot q,$$
$$0 \leqslant b \wedge p \leqslant q \Rightarrow b \cdot p \leqslant b \cdot q,$$

which, by means of a suitable change of variables, can be transformed into (i). ∎

Thus the axioms 7.5.1 and 7.5.2 may be equivalently replaced by the following law connecting the relation $\leqslant$ with the operations $+$ and $\cdot$:

Ax 7.5.5. $\quad a \leqslant b \wedge p \leqslant q \Rightarrow a \cdot q + b \cdot p \leqslant a \cdot p + b \cdot q.$[22]

Hence, to describe the ordered quasi-fields we do not need 7.5.2, it suffices to assume 7.5.1. Nevertheless, in application to geometry it is more convenient to characterize the ordered left quasi-fields by means of the axiom 7.5.5 instead of its particular case 7.5.1 (compare

[22] In a two-sided quasi-field (so, in particular, in a field), the axiom 7.5.5 can be expressed in terms of the positive cone P (comp. § 7.4): $x \in P \wedge y \in P \Rightarrow x \cdot y \in P$ (i.e. $P \cdot P \subseteq P$) (M.M).

the construction of the ordered quasi-field of scalars in an affine plane, § 8.2).

As a direct consequence of 7.5.5 we obtain[23]

Theorem 7.5.6. *Let* $(\mathfrak{F}, \leqslant)$ *be an ordered left quasi-field and let* $a \in F$. *If* $a > 0$ $(a < 0)$, *then the bijection* ${}^a\psi$ *is increasing (decreasing).*[24]

The discussion in § 6.1 (see theorem 6.1.1 together with the remark following it) combined with theorem 7.4.1 results in

Theorem 7.5.7. *Let* $\mathfrak{F} = (F, 0, 1, +, \cdot) \in \mathbf{QF}$. *If the group* $\mathfrak{F}_+ = (F, 0, +)$ *is orderable, then* $\mathfrak{F} \in \mathbf{QF}_{\left(\frac{1}{2}\right)}$.

We use the symbols

$$\mathbf{OQF}, \quad \mathbf{OF}, \quad \text{com}\,\mathbf{OF}$$

to denote the class of ordered strong left quasi-fields, ordered fields, and ordered commutative fields, respectively.
Of course

$$\mathbf{OQF} \supseteq \mathbf{OF} \supseteq \text{com}\,\mathbf{OF}.$$

As a corollary of 7.5.7 we get

Theorem 7.5.8. (i) $(\mathfrak{F}, \leqslant) \in \mathbf{OQF} \Rightarrow \mathfrak{F} \in \mathbf{QF}_{\left(\frac{1}{2}\right)}$.

(ii) $(\mathfrak{F}, \leqslant) \in \mathbf{OF} \Rightarrow \mathfrak{F} \in \mathbf{F}_{\left(\frac{1}{2}\right)}$.

(iii) $(\mathfrak{F}, \leqslant) \in \text{com}\,\mathbf{OF} \Rightarrow \mathfrak{F} \in \text{com}\,\mathbf{F}_{\left(\frac{1}{2}\right)}$.

Let us prove

Theorem 7.5.9. *Let* $(\mathfrak{F}, \leqslant)$ *be an ordered field. If the order* $\leqslant$ *is Archimedean in the group* $\mathfrak{F}_+$, *then the field* $\mathfrak{F}$ *is commutative.*

P r o o f.[25] For arbitrary $x \in F$ and $n \in N$ we define the product $x \cdot n$,

$$x \cdot 0 = 0, \quad x \cdot n = x \cdot (n-1) + x \quad \text{for} \quad n \in N - \{0\}.$$

It is easy to check (by induction) that

(1) $\quad x \cdot n = n \cdot x \quad$ for every $n \in N$ and $x \in F$.

---

[23] M.M.—to the end of § 7.5.

[24] See 0.6.1(ii).

[25] See Hilbert [50], p. 101.

Suppose $a \cdot b \neq b \cdot a$ for some $a, b \in F$. Without any restriction of generality of the proof we may assume

$$a > 0, \quad b > 0, \quad a \cdot b - b \cdot a > 0.$$

Let

(2) $$c = (a+b+1)^{-1} \cdot (a \cdot b - b \cdot a)$$

and

(3) $$d = \min(1, c).$$

By the axiom Arch there exist integers $m, n \geqslant 0$ such that

(4) $$\begin{aligned} m \cdot d < a \leqslant (m+1) \cdot d \\ n \cdot d < b \leqslant (n+1) \cdot d. \end{aligned}$$

From (1) and (4) it follows that

$$a \cdot b \leqslant m \cdot n \cdot d^2 + (m+n+1) \cdot d^2$$

and

$$b \cdot a > m \cdot n \cdot d^2,$$

whence

(5) $$a \cdot b - b \cdot a < (m+n+1) \cdot d^2.$$

The condition (5) combined with (3) and (4) implies

$$a \cdot b - b \cdot a < (a+b+1) \cdot c,$$

which contradicts (2). Hence $a \cdot b = b \cdot a$ for arbitrary $a, b \in F$. ∎

Theorem 7.5.10. *Let* $(\mathfrak{F}, \leqslant)$ *be an ordered left quasi-field. If the order* $\leqslant$ *is continuous, then* $\mathfrak{F}$ *is a field.*

P r o o f.[26] We have to prove that the multiplication is associative and right-distributive with respect to the addition:

(i) $$(u \cdot v) \cdot x = u \cdot (v \cdot x)$$

and

(ii) $$(u+v) \cdot x = u \cdot x + v \cdot x.$$

Obviously, we may assume that $u, v \neq 0$. By the left distributivity of the multiplication with respect to the addition, for arbitrary $u \in F - \{0\}$ the bijection ${}^u\psi$ is an automorphism of the group $\mathfrak{F}_+$.

[26] Compare Lenz [61], p. 103–105.

(i) Consider the bijections ${}^{u\cdot v}\psi$ and ${}^{u}\psi{}^{v}\psi$. If $u, v > 0$ or $u, v < 0$, then $u \cdot v > 0$ and, by 7.5.6, these bijections are automorphisms of the ordered group $(\mathfrak{F}_{+}, \leqslant)$. Evidently

$${}^{u\cdot v}\psi(1) = {}^{u}\psi{}^{v}\psi(1).$$

Since in a continuously ordered group for any a, b there is a unique automorphism carrying a on b,[27] the automorphism ${}^{u\cdot v}\psi$ coincides with ${}^{u}\psi{}^{v}\psi$,

$${}^{u\cdot v}\psi(x) = {}^{u}\psi{}^{v}\psi(x) \quad \text{for every } x \in F.$$

This yields (i) for $u \cdot v > 0$: In view of 0.3.2, this formula remains valid for $u > 0$ and $v < 0$, as well as for $u < 0$ and $v > 0$.

(ii) Consider the bijections ${}^{u+v}\psi$ and ${}^{u}\psi + {}^{v}\psi$. If $u, v > 0$, then $u+v > 0$ and, by the argument used in the proof of (i),

$$(2) \qquad {}^{u+v}\psi(x) = {}^{u}\psi(x) + {}^{v}\psi(x) \quad \text{for every } x \in F.$$

This yields (ii) for $u, v > 0$. Since (2) is symmetric with respect to the variables u and v, setting $u' = -u$, $v' = -v$ if $u, v < 0$ and $u' = u$, $v' = -v$ if $u > 0$, $v < 0$, $u+v > 0$, we easily check that (ii) remains valid for arbitrary $u, v \in F - \{0\}$. ∎

From 7.5.9 and 7.5.10 combined with 7.3.14 we derive

Theorem 7.5.11. *Let $(F, \leqslant)$ be an ordered left quasi-field. If the order $\leqslant$ is continuous, then $\mathfrak{F}$ is a commutative field.*

7.6. ORDERED MIDPOINT ALGEBRAS

Let us consider a midpoint algebra $\mathfrak{G} = (G, \oplus)$ and a betweenness relation $\boldsymbol{B}$ on the set G. The following sentences express some possible natural connections between the relation $\boldsymbol{B}$ and the operations $\oplus$ and $\ominus$.

Cor.	$\boldsymbol{B}(a\ a\oplus b\ b)$	(*correspondence*).
Inv.	$\boldsymbol{B}(pqr) \Rightarrow \boldsymbol{B}(a\ominus p\ a\ominus q\ a\ominus r)$	(*invariance*).
oInv.	$\boldsymbol{B}(pqr) \Rightarrow \boldsymbol{B}(p\ p\ominus q\ p\ominus r)$	(*outer invariance*).
iInv.	$\boldsymbol{B}(pqr) \Rightarrow \boldsymbol{B}(q\ominus p\ q\ q\ominus r)$	(*inner invariance*).

[27] See Lenz [61], p. 103, corollary.

The condition Cor says that the midpoint of a pair of points lies between these points. The condition Inv states that $\boldsymbol{B}$ is invariant under any reflection. The conditions oInv and iInv are particular cases of Inv. Of course, oInv and iInv can be formulated directly in terms of the original operation $\oplus$ as follows:

Inv′. $a = p\oplus p' = q\oplus q' = r\oplus r' \wedge \boldsymbol{B}(pqr) \Rightarrow \boldsymbol{B}(p'q'r')$,

oInv′. $p = q\oplus q' = r\oplus r' \wedge \boldsymbol{B}(pqr) \Rightarrow \boldsymbol{B}(pq'r')$,

iInv′. $p\oplus p' = q = r\oplus r' \wedge \boldsymbol{B}(pqr) \Rightarrow \boldsymbol{B}(p'qr')$.

The following inferences are based on the axioms for midpoint algebras (5.2.1–5.2.4) together with the definition of $\ominus$ (5.2.8) and the axioms for the betweenness relations (7.1.6–7.1.10).

Th 7.6.1. oInv $\mapsto$ Cor $\mapsto$ iInv.

P r o o f. Assume oInv. Suppose $\sim \boldsymbol{B}(a\; a\oplus b\; b)$ for some a and b; then

$$\boldsymbol{B}(a\oplus b\; a\; b) \quad \text{or} \quad \boldsymbol{B}(a\oplus b\; b\; a),$$

which, together with

$$a\oplus b = a\oplus b = b\oplus a,$$

by oInv′ yields

$$\boldsymbol{B}(a\oplus b\; a\; b) \wedge \boldsymbol{B}(a\oplus b\; b\; a).$$

Then $a = b = a\oplus b$, whence $\boldsymbol{B}(a\; a\oplus b\; b)$. Therefore oInv $\mapsto$ Cor.

Assume Cor. Since iInv holds for $p = q$ or $q = r$, we may in addition assume $p \neq q \neq r$. Let $\boldsymbol{B}(pqr)$. By Cor,

$$\boldsymbol{B}(p\; q\; q\ominus p) \quad \text{and} \quad \boldsymbol{B}(r\; q\; q\ominus r).$$

On the basis of the axioms for ordered sets we have the following inferences:

$$p \neq q \wedge \boldsymbol{B}(p\,q\,r) \wedge \boldsymbol{B}(p\; q\; q\ominus p) \mapsto \boldsymbol{B}(q\; r\; q\ominus p) \vee \boldsymbol{B}(q\; q\ominus p\; r),$$

$$q \neq r \wedge (\boldsymbol{B}(q r q\ominus p) \vee \boldsymbol{B}(q q\ominus p r)) \wedge \boldsymbol{B}(r q q\ominus r) \mapsto \boldsymbol{B}(q\ominus p\, q\, q\ominus r).$$

Therefore Cor $\mapsto$ iInv. ∎

Th 7.6.2. Inv $\leftrightarrows$ oInv.

P r o o f. The inference Inv $\mapsto$ oInv is obvious. Assume, conversely, oInv. In view of 7.6.1 we may also apply the law iInv. Clearly the condition Inv is satisfied for $p = q$ or $q = r$, and, by iInv, for $a = q$ as well. So, assume

$$p \neq q \neq r \quad \text{and} \quad a \neq q$$

and let $\boldsymbol{B}(pqr)$. Then

$$\boldsymbol{B}(apqr) \vee \boldsymbol{B}(paqr) \vee \boldsymbol{B}(pqar) \vee \boldsymbol{B}(pqra),$$

whence by oInv and iInv

$$\boldsymbol{B}(a\, a\ominus p\, a\ominus q) \wedge \boldsymbol{B}(a\, a\ominus q\, a\ominus r)$$

or

$$\boldsymbol{B}(a\ominus p\, a\, a\ominus q) \wedge \boldsymbol{B}(a\, a\ominus q\, a\ominus r)$$

or

$$\boldsymbol{B}(a\ominus r\, a\, a\ominus q) \wedge \boldsymbol{B}(a\, a\ominus q\, a\ominus p)$$

or

$$\boldsymbol{B}(a\, a\ominus r\, a\ominus q) \wedge \boldsymbol{B}(a\, a\ominus q\, a\ominus p).$$

Each of these four cases immediately yields $\boldsymbol{B}(a\ominus p\, a\ominus q\, a\ominus r)$. Thus oInv $\models\!\!\Rightarrow$ Inv. ■

The structure $(\mathfrak{G}, \boldsymbol{B}) = (G, \oplus, \boldsymbol{B})$ consisting of a midpoint algebra $\mathfrak{G} = (G, \oplus)$ and a betweenness relation $\boldsymbol{B}$ on G is said to be an *ordered midpoint algebra* whenever the relation $\boldsymbol{B}$ is invariant under every reflection of $\mathfrak{G}$, i.e. the general invariance law Inv is satisfied in $(G, \boldsymbol{B})$. Then $\boldsymbol{B}$ is referred to as a *(linear) order in* $\mathfrak{G}$.

The axiom system for the ordered midpoint algebras can be a little simplified. The set G, as the universe of a midpoint algebra, has cardinality $\neq 4$,[28] and thus the axiom 7.1.10 may be omitted (see § 7.2); in view of 7.6.2, the invariance axiom Inv can be equivalently replaced by its particular case oInv. In conclusion, the set of sentences

$$\{5.2.1\text{–}5.2.4\} \cup \{7.1.6\text{–}7.1.9\} \cup \{\text{oInv}\}$$

is an axiom system for ordered midpoint algebras.

The notion of ordered midpoint algebra is closely related to the notion of ordered group. As known, the midpoint algebras are the arithmetic mean algebras for groups belonging to $\mathbf{G}_{\left(\frac{1}{2}\right)}$ (see § 5.3). So, let us consider any group $\mathfrak{G} = (G, 0, +) \in \mathbf{G}_{\left(\frac{1}{2}\right)}$ together with

[28] The simple proof of this fact is left to the reader (M.M.).

its arithmetic mean algebra $(G, \oplus)$, and any betweenness relation $\boldsymbol{B}$ on G together with the pair of linear orders adjoint to $\boldsymbol{B}$, $\Psi(\boldsymbol{B}) = \{\leqslant, \geqslant\}$. By the definitional equivalence of the structures

$$(G, 0, \oplus) \quad \text{and} \quad (G, 0, +)$$

and of the structures

$$(G, \boldsymbol{B}) \quad \text{and} \quad (G, \{\leqslant, \geqslant\}),$$

the structures

$$(G, 0, \oplus, \boldsymbol{B}) \quad \text{and} \quad (G, 0, +, \{\leqslant, \geqslant\})$$

have common definitional expansion. We are going to prove that in the theory of this expansion the invariance law Inv, which characterizes the structure $(G, \oplus, \boldsymbol{B})$ as an ordered midpoint algebra, is equivalent to the monotony law (7.3.0)

Mon. $p \leqslant q \Rightarrow p+a \leqslant q+a,$

which characterizes each of the structures $(G, 0, +, \leqslant)$ and $(G, 0, +, \geqslant)$ as an ordered group. Indeed, consider the auxiliary law

Refl. $p \leqslant q \Rightarrow a\ominus p \geqslant a\ominus q,$

which states that any reflection permutes the pair $\{\leqslant, \geqslant\}$. Let us show that Refl is equivalent to each of Inv and Mon.

Th 7.6.3. Refl $\rightleftarrows$ Inv.

P r o o f. The inference Refl $\Rightarrow$ Inv follows immediately from the definition of $\boldsymbol{B}$ in terms of $\{\leqslant, \geqslant\}$, (7.1.17). Assume conversely Inv; in view of 7.6.1 and 7.6.2 we can apply Cor. Let us reformulate the sentence Refl replacing the variable "a" by "b" (this will make it easier to use in the proof the definition of $\{\leqslant, \geqslant\}$ in terms of $\boldsymbol{B}$ (7.1.18)). We obtain

Refl′. $p \leqslant q \Rightarrow b\ominus p \geqslant b\ominus q.$

Of course, Refl′ is satisfied if b is the unique element of the universe and then $p = q = b$. Thus, we may assume that there exists a point a different from b. Let $c = b\ominus a$. Then $\neq (abc)$ and $b = a\oplus c$, whence, by Cor, we get $\boldsymbol{B}(abc)$. Without any restriction of generality of the proof we may assume that

$$\Psi(\boldsymbol{B}; a, b, c) = \leqslant.$$

Then

$$\Psi(\boldsymbol{B}; c, b, a) = \geqslant$$

and, by 7.1.18 combined with Inv,

$$p \leqslant q \mapsto b \ominus p \geqslant b \ominus q.$$

Hence Inv $\mapsto$ Refl. ∎

Th 7.6.4. Mon $\leftrightarrows$ Refl.

Proof. From 5.2.8 and 5.1.6 it follows that

$$a \ominus p = a + a - p,$$

$$p + a = \frac{1}{2} a \ominus (0 \ominus p).$$

In view of 7.3.1, the first of these two formulae gives the inference Mon $\mapsto$ Refl. By the second one, the bijection σ^a is a composition of two reflections, with respect to 0 and with respect to $\frac{1}{2}a$. This immediately yields the inference Refl $\mapsto$ Mon. ∎

By 7.6.3 and 7.6.4 the laws Inv and Mon are equivalent. Therefore the structures $(G, 0, \oplus, \boldsymbol{B})$ and $(G, 0, +, \{\leqslant, \geqslant\})$ are definitionally equivalent. Thus the structure $(G, \oplus, \boldsymbol{B})$ is an ordered midpoint algebra if and only if each (actually, if one) of the structures $(G, 0, +, \leqslant)$ and $(G, 0, +, \geqslant)$ is an ordered group. We say that the ordered midpoint algebra $(G, \oplus, \boldsymbol{B})$ is adjoint to the ordered group $(G, 0, +, \leqslant)$ (to $(G, 0, +, \geqslant)$).

An order $\boldsymbol{B}$ in a midpoint algebra $(G, \oplus)$ is said to be ***Archimedean***[29] if the structure $(G, \oplus, \boldsymbol{B})$ satisfies the following additional axiom

Ax Arch′. $p_0 \neq p_1 \wedge \boldsymbol{B}(p_0 p_1 b) \wedge \forall i \in N \; p_i \oplus p_{i+2} = p_{i+1}$
$\Rightarrow \exists n \in N \; \boldsymbol{B}(p_0 b p_n).$

Let us prove

Theorem 7.6.5. *Let $(G, 0, +, \leqslant)$ be an ordered group and let $(G, \oplus, \boldsymbol{B})$ be the ordered midpoint algebra adjoint to $(G, 0, +, \leqslant)$. The order $\leqslant$ is Archimedean if and only if the relation $\boldsymbol{B}$ is Archimedean; i.e.*

$$\text{Arch} \leftrightarrows \text{Arch}'.$$

[29] M.M.—to the end of § 7.6.

P r o o f. Arch $\Rightarrow$ Arch'. Assume Arch. Let

(1) $p_0 \neq p_1 \wedge B(p_0 p_1 b)$

and

(2) $p_i \oplus p_{i+2} = p_{i+1}$ for $i = 0, 1, \ldots$

Without any restriction of generality we may strengthen the assumption (1) admitting

(3) $0 < p_0 < p_1 < b.$

From (2) it follows that

(4) $p_{i+2} - p_{i+1} = p_{i+1} - p_i$ for $i = 0, 1, \ldots$

Let

(5) $a = p_1 - p_0.$

By Arch from (3) and (5) it follows that there is a natural number n satisfying the condition

(6) $n \cdot a \geqslant b,$

which, by (4) and (5), is equivalent to

(6') $p_n - p_0 \geqslant b.$

From (3) and (6') it follows that $p_0 < b < p_n$, whence

$$B(p_0 b p_n).$$

Arch' $\Rightarrow$ Arch. Assume Arch' and let

(7) $a, b > 0.$

The law Arch obviously holds for $a > b$, so we can assume

(8) $a \leqslant b.$

Let

(9) $p_i = i \cdot a$ for $i \in N.$

From (7)–(9) it follows that $p_0 \neq p_1 \wedge B(p_0 p_1 b) \wedge p_i \oplus p_{i+2} = p_{i+1}$, thus, by Arch', there is a number n satisfying the condition

(10) $B(p_0 b p_n)$,

which implies $b \leqslant p_n$, and so $n \cdot a \geqslant b$. ■

Let us notice that in the proof of the inference Arch′ $\Rightarrow$ Arch we have not used the law Mon. This will be essential in § 7.7.

7.7. ORDERINGS COMPATIBLE WITH A MIDPOINT OPERATION

In the last section, basing on the axioms for midpoint algebras, the definition of $\ominus$, and the axioms for betweenness relations, we established the following two inferences:

$$\text{oInv} \Rightarrow \text{Cor} \Rightarrow \text{iInv}.$$

What about the converse inferences? If Cor implied oInv, then in the axiom system for ordered midpoint algebras the sentence oInv could be replaced by Cor, which is simpler than oInv as regards form and meaning. Unfortunately, this is not the case.

Th 7.7.1. $\sim$ (oInv $\Leftarrow$ Cor).

Proof. We have to construct a midpoint algebra $\mathfrak{G} = (G, \oplus)$ and a betweenness relation $\boldsymbol{B}$ on G such that the structure $(G, \boldsymbol{B})$ satisfies Cor but not oInv. Let us start with the additive group of rational numbers $\mathfrak{Q} = (Q, 0, +)$ with its natural ordering $\leqslant$. Clearly $\mathfrak{Q} \in \mathbf{G}_{\left(\frac{1}{2}\right)}$. Cartesian square $\mathfrak{Q}^2 = (Q^2, (0, 0), +)$ of $\mathfrak{Q}$ also belongs to $\mathbf{G}_{\left(\frac{1}{2}\right)}$. Let $\leqslant$ be the lexicographical order in Q^2 generated by $\leqslant$. It may be described as follows. The set Q^2 is the union of sets

$$Q_v = \{(v, w) : w \in Q\} \quad \text{for} \quad v \in Q.$$

Notice that Q_v are segments of the ordered set $(Q^2, \leqslant)$. We have

$$Q_{v_1} \leqslant Q_{v_2} \Leftrightarrow v_1 \leqslant v_2$$

and, for any segment Q_v,

$$(v, w_1) \leqslant (v, w_2) \Leftrightarrow w_1 \leqslant w_2.$$

Evidently the structure $(\mathfrak{Q}^2, \leqslant)$ is an ordered group.

We must now "spoil" a little the relation $\leqslant$. Let $\leqslant'$ differ from $\leqslant$ only on the segment Q_{-1}, where the order $\leqslant$ is replaced by its converse, $\geqslant$. This new relation $\leqslant'$ is again an ordering of Q^2. Moreover, for any two points $p = (v_1, w_1)$, $q = (v_2, w_2) \in Q^2$ we have

$$p \leqslant' q \Rightarrow p \leqslant' \frac{1}{2}(p+q) \leqslant' q. \tag{1}$$

Indeed, take three points, p, q, and $\frac{1}{2}(p+q) = \left(\frac{1}{2}(v_1+v_2), \frac{1}{2}(w_1+w_2)\right)$, and notice that either no one of them belongs to Q_{-1}, or only one, or all three of them belong to Q_{-1}. Thus in the formula (1) the relation $\leqslant'$ can be equivalently replaced by $\leqslant$ or $\geqslant$.

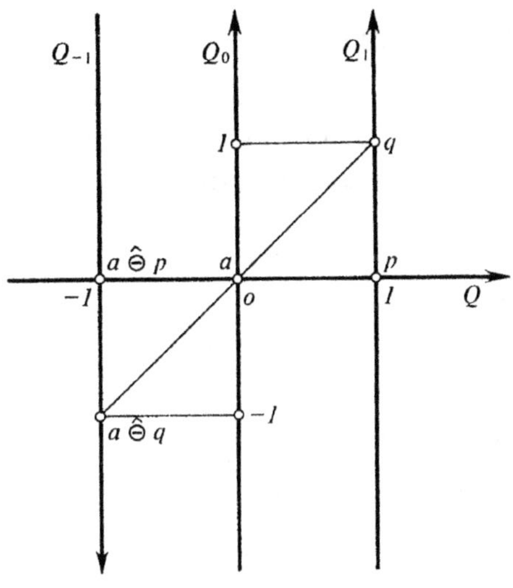

Fig. 54

Let us now pass from the group $\mathfrak{Q}^2$ and the order $\leqslant'$ in Q^2 to the arithmetic mean algebra $\mathfrak{G} = (Q^2, \oplus)$ and the betweenness relation $\boldsymbol{B}'$ adjoint to $\leqslant'$. By (1), the structure $(\mathfrak{G}, \boldsymbol{B}')$ satisfies Cor. At the same time this structure does not satisfy oInv. Indeed, if

$$a = (0, 0), \quad p = (1, 0), \quad \text{and} \quad q = (1, 1),$$

then

$$a \ominus p = a + a - p = (-1, 0) \quad \text{and} \quad a \ominus q = a + a - q = (-1, -1),$$

and thus (see fig. 54)

$$\boldsymbol{B}'(apq) \quad \text{and} \quad \boldsymbol{B}'(a\ominus p\ a\ominus q\ a),$$

whence

$$\boldsymbol{B}'(apq) \quad \text{and} \quad \sim \boldsymbol{B}'(a\ a\ominus p\ a\ominus q).$$

As a result, the law oInv is independent of Cor; the structure $(\mathfrak{G}, \boldsymbol{B}')$ is a model of independence. ∎

Let us notice that the relation $\boldsymbol{B}'$ is non-Archimedean in the midpoint algebra $\mathfrak{G}$. Indeed, the order $\leqslant$, as a lexicographic one, is non-Archimedean in the group $\mathfrak{Q}^2$ and consequently the order $\leqslant'$ and the adjoint relation $\boldsymbol{B}'$ are non-Archimedean as well.[30]

Is this essential for independence models that $\boldsymbol{B}'$ is non-Archimedean? The answer is affirmative,

Th 7.7.2. oInv $\Leftarrow_{\mathrm{I}}$ Arch′ $\wedge$ Cor.

Proof. Let $\mathfrak{G} = (G, \oplus)$ be a midpoint algebra and let $\boldsymbol{B}$ be a betweenness relation on G. Let us assume that $(\mathfrak{G}, \boldsymbol{B})$ satisfies Arch′ and Cor; then, by 7.6.1, it satisfies iInv as well.

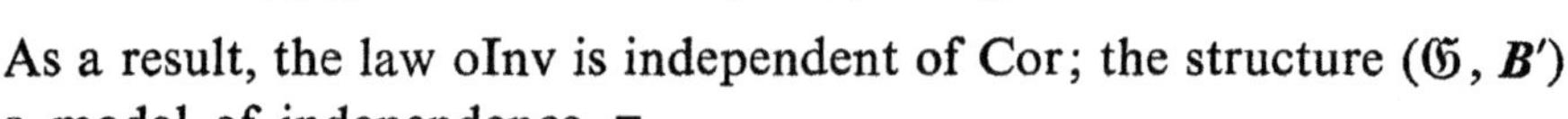

Fig. 55

Take $a, p, q \in G$. Assume

(2) $\neq (apq)$ and $\boldsymbol{B}(apq)$.

Suppose

(3) $\boldsymbol{B}(a\ a\ominus p\ a\ominus q)$.

In view of (2) the law iInv excludes the case $\boldsymbol{B}(a\ominus p\ a\ a\ominus q)$, and thus the condition (3) implies

(4) $\boldsymbol{B}(a\ a\ominus q\ a\ominus p)$.

Let us now consider two sequences of points (see fig. 55), the sequence $r_0, r_1, \ldots$ defined by the conditions

(5) $r_0 = p, \quad r_1 = q, \quad r_{i+2} = r_{i+1}\ominus r_i \quad$ for $\quad i = 0, 1, \ldots$

and the sequence $s_0, s_1, \ldots$ defined by the condition

(6) $s_i = a\ominus r_i \quad$ for $\quad i = 0, 1, \ldots$

---

[30] See Theorem 7.6.5 together with the remark following it. (M.M.).

From (5), by Cor, it follows that

$$\boldsymbol{B}(r_i r_{i+1} r_{i+2}) \quad \text{for} \quad i = 0, 1, \ldots,$$

which, together with (2), implies

(7) $\boldsymbol{B}(a r_0 r_i) \quad \text{for} \quad i = 0, 1, \ldots$

In turn, the formula (4) yields $\boldsymbol{B}(s_0 s_1 a)$, and thus by Arch′ we obtain

$$\boldsymbol{B}(s_0 a s_n)$$

for a sufficiently large n. But this implies, by (6) and iInv,

$$\boldsymbol{B}(r_0 a r_n)$$

contrary to (7). Thus the supposition (3) leads to contradiction, whence $\boldsymbol{B}(a\; a\ominus p\; a\ominus q)$. Therefore the structure $(\mathfrak{G}, \boldsymbol{B})$ satisfies oInv. ■

Let $\mathfrak{G} = (G, \oplus)$ be a midpoint algebra and let $\boldsymbol{B}$ be a betweenness relation on G. If $(G, \boldsymbol{B})$ satisfies Cor, we say that the relation $\boldsymbol{B}$ *is compatible with the midpoint operation.*

In view of 7.6.1, if $\boldsymbol{B}$ is an order in $\mathfrak{G} = (G, \oplus)$, then $\boldsymbol{B}$ is compatible with $\oplus$. On the other hand, by 7.7.1, the compatibility of $\boldsymbol{B}$ with the midpoint operation is not sufficient for $\boldsymbol{B}$ to be an order in $\mathfrak{G}$, unless $\boldsymbol{B}$ is Archimedean (or, in particular, continuous).

The answer to the second question, whether iInv implies Cor, is also negative:

Th 7.7.3.[31] $\sim$ (Cor $\Leftarrow_{\mathrm{I}}$ iInv).

Proof. Let $G = \{p_0, p_1, p_2\}$. We define the operation $\oplus$ and the relation $\boldsymbol{B}$ as follows:

$$p_i \oplus p_j = p_k \Leftrightarrow i = j = k \vee \neq (i, j, k);$$

$$\boldsymbol{B}(p_i p_j p_k) \Leftrightarrow i \leqslant j \leqslant k \vee k \geqslant j \geqslant i.$$

It is easy to check that $\oplus$ is a midpoint operation and $\boldsymbol{B}$ is a betweenness relation on G; moreover iInv is satisfied in $(G, \oplus, \boldsymbol{B})$. On the other hand

$$\boldsymbol{B}(p_0\; p_1\; p_0 \oplus p_1),$$

whence Cor is not satisfied. Thus the law Cor is independent of iInv; the structure $(G, \oplus, \boldsymbol{B})$ is an independence model. ■

Let us notice that this relation $\boldsymbol{B}$ is Archimedean in $(G, \oplus)$.

---

[31] Due to L. Dubikajtis (M.M.).

CHAPTER 8

ORDERED AFFINE PLANES

8.1. THE BETWEENNESS RELATION ON AN AFFINE PLANE AND THE PASCH AXIOM

Let us consider a parallelity plane $\mathfrak{A} = (\boldsymbol{S}, \|)$ and a betweenness relation $\boldsymbol{B} \subseteq \boldsymbol{S} \times \boldsymbol{S} \times \boldsymbol{S}$. The relation $\boldsymbol{B}$ is said to be a *betweenness relation on the plane* $\mathfrak{A}$ whenever the structure $(\mathfrak{A}, \boldsymbol{B})$ satisfies the following two conditions:

(i) $\boldsymbol{B}$ is a linear relation, i.e. $\boldsymbol{B} \subseteq \boldsymbol{L}$,

(ii) $\boldsymbol{B}$ (more precisely $\boldsymbol{B}|K$) is a betweenness relation on every line K on the plane $\mathfrak{A}$.

The affine plane $\mathfrak{A}$ is described by the elementary axiom system AX 2.2. The elementary form of (i) is

Ax 8.1.0. $\boldsymbol{B}(abc) \Rightarrow \boldsymbol{L}(abc)$.

The condition (ii) can be equivalently replaced by the linear modification of the (elementary) axioms 7.1.6–7.1.10. In fact, only the axiom 7.1.6 has to be modified; we change it into

Ax 8.1.1. $\boldsymbol{L}(abc) \Rightarrow \boldsymbol{B}(abc) \vee \boldsymbol{B}(bca) \vee \boldsymbol{B}(cab)$.

The remaining axioms, 7.1.7–7.1.10, need not be changed. Indeed, each of them is an implication, from the antecedent of which it follows that (by 8.1.0, 1.1.1′, 1.1.2′, and, for 7.1.9, also by 7.1.7) all the points involved belong to a line K.

The axioms 8.1.0 and 8.1.1 yield

Th 8.1.2. $\boldsymbol{L}(abc) \Leftrightarrow \boldsymbol{B}(abc) \vee \boldsymbol{B}(bca) \vee \boldsymbol{B}(cab)$.

Thus in the structure $(\mathfrak{A}, \boldsymbol{B})$ collinearity can be simply defined by means of the betweenness relation.

Let us notice that in the axiom system for $(\mathfrak{A}, \boldsymbol{B})$ the axiom 7.1.7 is superflous, since it follows (by 2.2.1, 2.2.2, and 2.2.4) from the remaining axioms for the betweenness relation. Indeed, without use of 7.1.7 we prove by turns

Th 8.1.3. $\boldsymbol{B}(aaa)$.

Proof. By 8.1.1,

$$\boldsymbol{L}(aaa) \mapsto \boldsymbol{B}(aaa). \blacksquare$$

Th 8.1.4. $\boldsymbol{B}(aab)$.

Proof. By 8.1.3 we have $\boldsymbol{B}(aaa)$. Thus, by 8.1.1, 7.1.9, and 7.1.8

$$\boldsymbol{L}(aab) \mapsto \boldsymbol{B}(aab) \vee \boldsymbol{B}(aba) \vee \boldsymbol{B}(baa),$$
$$\boldsymbol{B}(aba) \wedge \boldsymbol{B}(aaa) \mapsto \boldsymbol{B}(baa),$$
$$\boldsymbol{B}(baa) \mapsto \boldsymbol{B}(aab). \blacksquare$$

Th 8.1.5. (7.1.7) $\boldsymbol{B}(aba) \Rightarrow a = b$.

Proof. Assume $\boldsymbol{B}(aba)$ and let $a \neq b$. Then there is a point c such that

(1) $\sim \boldsymbol{L}(abc)$.

But 8.1.4 yields $\boldsymbol{B}(aac)$, and thus, by 7.1.9 and 8.1.0,

$$\boldsymbol{B}(aba) \wedge \boldsymbol{B}(aac) \mapsto \boldsymbol{B}(bac) \mapsto \boldsymbol{L}(bac),$$

which contradicts the condition (1). Hence $a = b$. $\blacksquare$

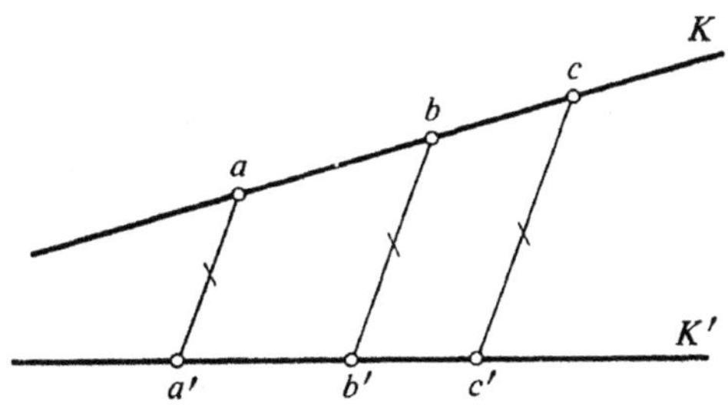

Fig. 56

Let $\mathfrak{A} = (\boldsymbol{S}, \|)$ be a parallelity plane and let $\boldsymbol{B}$ be a betweennes relation on $\mathfrak{A}$. Consider two distinct lines K and K' on $\mathfrak{A}$. The following three sentences describe some possible natural connections between the relations $\boldsymbol{B}|K$ and $\boldsymbol{B}|K'$:

Pasch. $K \neq K' \wedge a, b, c \in K \wedge a', b', c' \in K' \wedge aa' \| bb' \| cc' \wedge \boldsymbol{B}(abc) \Rightarrow \boldsymbol{B}(a'b'c')$ (*invariance*),

oPasch. $\sim L(abb') \wedge L(abc) \wedge L(ab'c') \wedge bb' \| cc' \wedge \boldsymbol{B}(abc) \Rightarrow \boldsymbol{B}(ab'c')$ (*outer invariance*),

iPasch. $\sim L(aa'b) \wedge L(abc) \wedge L(a'bc') \wedge aa' \| cc' \wedge \boldsymbol{B}(abc) \Rightarrow \boldsymbol{B}(a'bc')$ (*inner invariance*).

The sentence Pasch states that the relation $\boldsymbol{B}$ is preserved by an arbitrary parallel projection (fig. 56). The sentences oPasch (fig. 57) and iPasch (fig. 58) are particular cases of Pasch. Of course in both

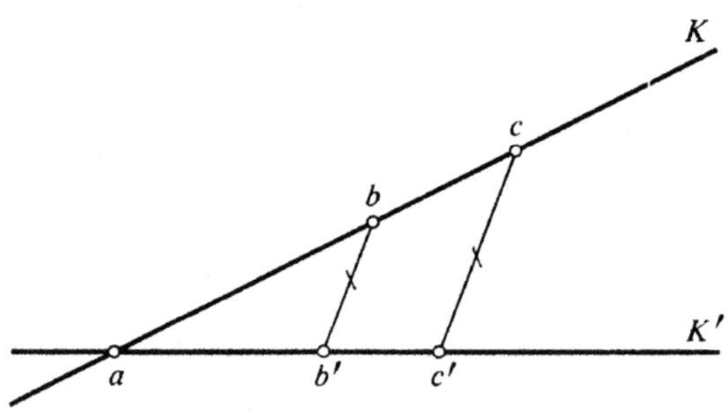

Fig. 57

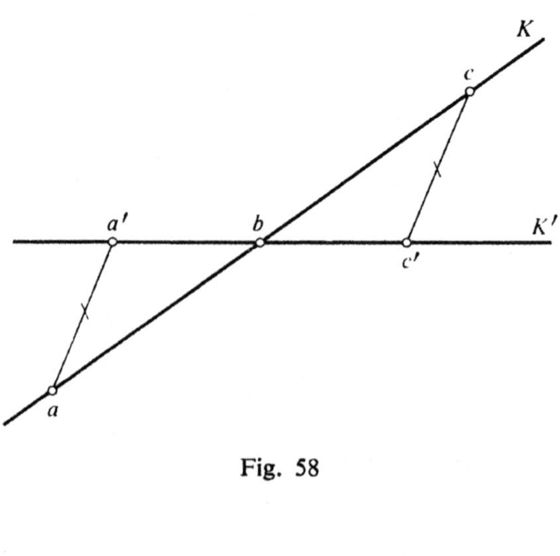

Fig. 58

of them the formula $L(abc)$ may be omitted. Notice that the assumption of the law Pasch may be strengthened by the additional condition $a \neq b \neq c$, because in each of the cases $a = b$ and $b = c$ this law becomes a (trivial) consequence of the axioms for $(\mathfrak{A}, \boldsymbol{B})$. By the same reason the assumptions of oPasch and iPasch may be strengthened by the additional condition $b \neq c$.

On the basis of the axioms for $(\mathfrak{A}, \boldsymbol{B})$, the laws Pasch, oPasch, and iPasch turn out to be equivalent:

Th 8.1.6. oPasch $\overset{\mid\Rightarrow}{\Leftarrow\mid}$ iPasch.

P r o o f. $\mid\Rightarrow$. Let the assumption of iPasch be satisfied, i.e.

$$\sim \boldsymbol{L}(aa'b), \quad \boldsymbol{B}(abc), \quad \boldsymbol{L}(a'bc'), \quad aa' \| cc'$$

and let $b \neq c$ (see fig. 57). Then

$$\boldsymbol{L}(cc'b).$$

Suppose $\sim \boldsymbol{B}(a'bc')$, i.e.

$$\boldsymbol{B}(ba'c') \vee \boldsymbol{B}(bc'a').$$

By the law oPasch, we get

$$\boldsymbol{B}(bac) \vee \boldsymbol{B}(bca),$$

which, combined with the assumption $\boldsymbol{B}(abc)$, yields $\boldsymbol{B}(a'bc')$.

In conclusion oPasch $\mid\Rightarrow$ iPasch.

$\Leftarrow\mid$. Let the assumption of oPasch be satisfied, i.e.

$$\sim \boldsymbol{L}(abb'), \ \boldsymbol{B}(abc), \ \boldsymbol{L}(ab'c'), \text{ and } bb' \| cc',$$

and let $b \neq c$ (see fig. 59). Then

$$\sim \boldsymbol{L}(c'bb').$$

Take the line M such that $a \in M$ and $M \| [bb']$. The line $[c'b]$ intersects M in some point a'. Obviously $\sim \boldsymbol{L}(aa'b)$ and thus, by iPasch, from $\boldsymbol{B}(abc)$ it follows that

$$\boldsymbol{B}(a'bc').$$

Suppose now $\sim \boldsymbol{B}(ab'c')$, i.e.

$$\boldsymbol{B}(b'ac') \vee \boldsymbol{B}(b'c'a).$$

By the law iPasch we get

$$\boldsymbol{B}(bac) \vee \boldsymbol{B}(bc'a'),$$

which, combined with $\boldsymbol{B}(abc)$ and $\boldsymbol{B}(a'bc')$, yields $\boldsymbol{B}(ab'c')$. In conclusion iPasch $\mid\!\Rightarrow$ oPasch. ■

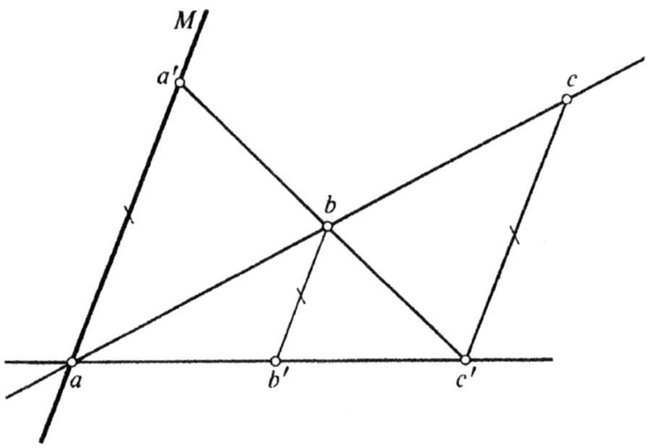

Fig. 59

Th 8.1.7. Pasch $\mid\!\rightleftarrows$ oPasch.

P r o o f. The inference Pasch $\mid\!\Rightarrow$ oPasch is obvious. Assume, conversely, oPasch. By 8.1.6 we can also apply the law iPasch. Let the assumption of Pasch be satisfied, i.e.

$$K \neq K', \quad a, b, c \in K, \quad a'b'c' \in K', \quad aa' \| bb' \| cc',$$
$$\text{and} \quad \boldsymbol{B}(abc),$$

and let $a \neq b \neq c$. If $b = b'$ then, by iPasch, $\boldsymbol{B}(a'b'c')$. Assume in turn $b \neq b'$ (see Fig. 60) and let $[ac'] \cap [bb'] = \{b''\}$ (in particular, if $a = a'$

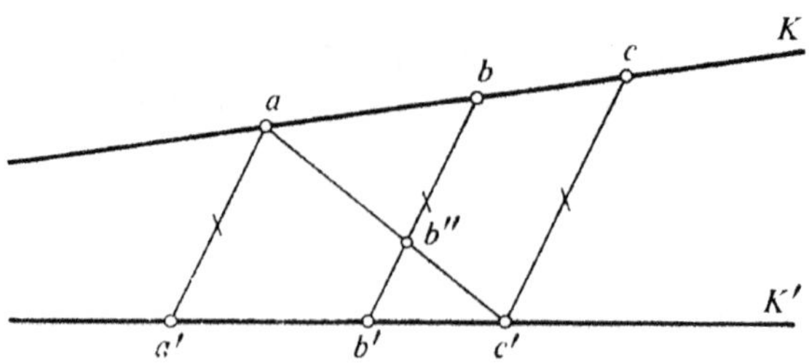

Fig. 60

then $b'' = b'$, and if $c = c'$ then $b'' = b$). Then, by Pasch, $\boldsymbol{B}(abc)$ implies by turns $\boldsymbol{B}(ab''c')$ and $\boldsymbol{B}(a'b'c')$. Therefore, oPasch $\models\!\!\Rightarrow$ Pasch. ■

It remains to compare these three invariance laws (referred here to as the Pasch laws) with the classical Pasch theorem, which states (in common parlance) the following:

Consider a triangle abc (so, $\sim \boldsymbol{L}(abc)$) and a line $K \not\ni a, b, c$. If K intersects the side ab of abc (i.e. the point of intersection of the lines K and $[ab]$ lies between a and b), then K intersects also at least one of the other sides of this triangle, ac or bc.

This sentence was admitted by Hilbert as an axiom of Euclidean geometry. Let us show that it is equivalent to each one of the above three laws, Pasch, oPasch, and iPasch.

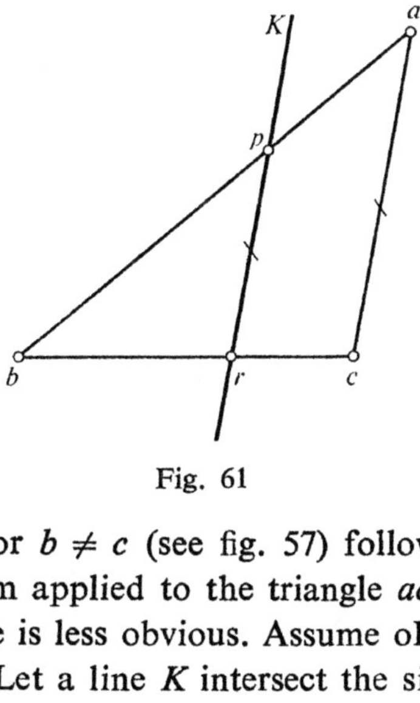

Fig. 61

The law oPasch for $b \neq c$ (see fig. 57) follows directly from the classical Pasch theorem applied to the triangle acc' and the line $[bb']$. The converse inference is less obvious. Assume oPasch; then, by 8.1.6, iPasch holds as well. Let a line K intersect the side ab of the triangle abc at some point p; then

$$\boldsymbol{B}(apb).$$

If $K \| [ac]$ (see fig. 61), then K intersects the line $[bc]$ at some point r, and, by oPasch, $\boldsymbol{B}(apb)$ implies $\boldsymbol{B}(brc)$, i.e. r belongs to the side bc. Analogically, if $K \| [bc]$, then K intersects the side ac of the triangle abc. Assume now that $\sim (K \| [ac])$ and $\sim (K \| [bc])$. Consider the line M determined by the conditions $a \in M$ and $M \| K$, and let

$$K \cap [ac] = \{q\}, \quad K \cap [bc] = \{r\}, \quad \text{and} \quad M \cap [bc] = \{s\}.$$

By oPasch $\boldsymbol{B}(apb)$ implies

$$\boldsymbol{B}(brs).$$

Now, if the line K does not intersect the side ac, then $\boldsymbol{B}(qac)$ or $\boldsymbol{B}(acq)$. But, by oPasch, $\boldsymbol{B}(qac)$ (see fig. 62) implies $\boldsymbol{B}(rsc)$, and, by iPasch, $\boldsymbol{B}(acq)$ (see fig. 63) implies $\boldsymbol{B}(rcs)$; this together with $\boldsymbol{B}(brs)$ and $r \neq s$ yields in both cases $\boldsymbol{B}(brc)$. Hence, the line K intersects the side bc. This completes the proof of the classical Pasch theorem.

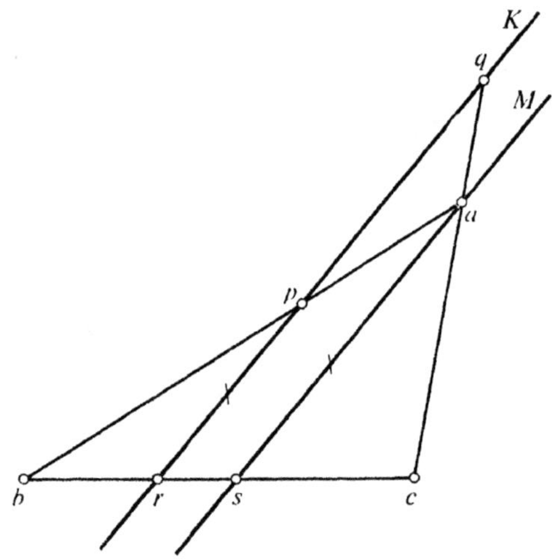

Fig. 62

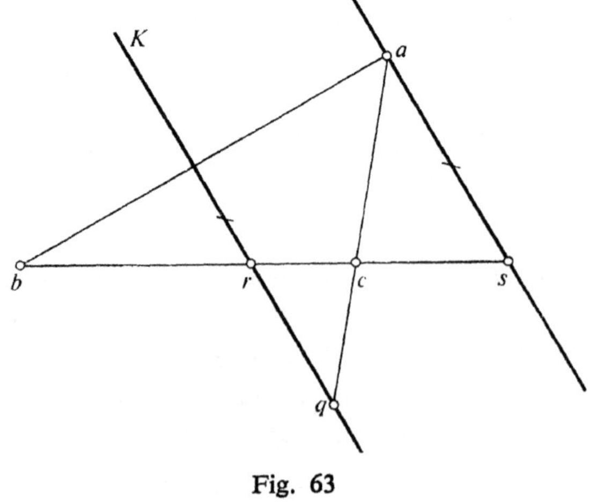

Fig. 63

The established equivalence justifies the use of the name Pasch for the invariance laws. However, it is easy to see from the proof of this equivalence, that the classical Pasch theorem says actually more than oPasch; but, the additional facts described by this theorem clearly follow already from the axioms for $(\mathfrak{A}, \boldsymbol{B})$ and thus do not need to be postulated. Moreover, oPasch is a universal sentence, while the classical Pasch theorem, when written in the formal language, is a conditional existential sentence. So, it is reasonable to replace the Hilbert form of Pasch axiom by oPasch.

8.2. ORDERED TRANSLATION PLANES[1]

Let $\mathfrak{A} = (\boldsymbol{S}, \|)$ be a translation plane. Assume that

(i) every line K on $\mathfrak{A}$ consists of at least three points.

Let $\boldsymbol{B}$ be a betweenness relation on $\mathfrak{A}$. Then the relation $\boldsymbol{B}|K$ is an ordering of K. Assume, moreover, that

(ii) every parallel projection of K onto a line K' different from K carries $\boldsymbol{B}|K$ onto $\boldsymbol{B}|K'$,

i.e. the structure $(\mathfrak{A}, \boldsymbol{B})$ satisfies the law Pasch. Then we say that the relation $\boldsymbol{B}$ is an ordering of the plane $\mathfrak{A}$. The structure $(\mathfrak{A}, \boldsymbol{B}) = (\boldsymbol{S}, \|, \boldsymbol{B})$ is then referred to as an ordered translation plane.

The condition (i) may be expressed by an elementary axiom

Ax 8.2.0. $\forall ab\, \exists c\, (a, b \neq c \wedge \boldsymbol{L}(abc))$.

By 8.1.6 and 8.1.7, the law Pasch, which express the condition (ii), may be equivalently replaced by any one of two elementary laws, oPasch and iPasch. We choose oPasch as an axiom. Hence the set of sentences

$$\{8.2.0\} \cup \mathrm{AX}\ 2.2 \cup \{\mathrm{des}\} \cup \{7.1.8\text{–}7.1.10\} \cup \{8.1.0,\ 8.1.1,\ \mathrm{oPasch}\}$$

is an axiom system for the ordered translation planes.

Let $(\mathfrak{A}, \boldsymbol{B})$ be an ordered translation plane. Let us choose parameters of coordinatization, $(o, 1, e)$, (compare § 4.1), and consider the left quasi-field of scalars, $\mathfrak{F}(o, 1, e) = (F, o, 1, +, \cdot)$. Let $\{\leqslant, \geqslant\}$ be the pair of linear orders in F, adjoint to the relation $\boldsymbol{B}|F$.

First of all we shall establish the connection between the pair $\{\leqslant, \geqslant\}$ and the additive group $\mathfrak{F}_+ = (F, o, +)$ of the quasi-field $\mathfrak{F}(o, 1, e)$.

---

[1] M.M.—editorial changes.

Consider the bijections $\sigma^u : F \to F$ and $\mu : F \to F$ defined by the formulae 0.6.1(i) and 0.6.2. We have

$$\sigma^u(x) = x+u \quad \text{for} \quad u \in F, \quad \mu(x) = -x.$$

By 4.3.5, each of the functions σ^u is the composition of three parallel projections in $\mathfrak{A}$. This also concerns the function μ (see fig. 64),

$$\mu : F \underset{\|E}{\to} I \underset{\|F}{\to} E \underset{\|I}{\to} F,$$
$$x \mapsto p_1 \mapsto p_2 \mapsto -x.$$

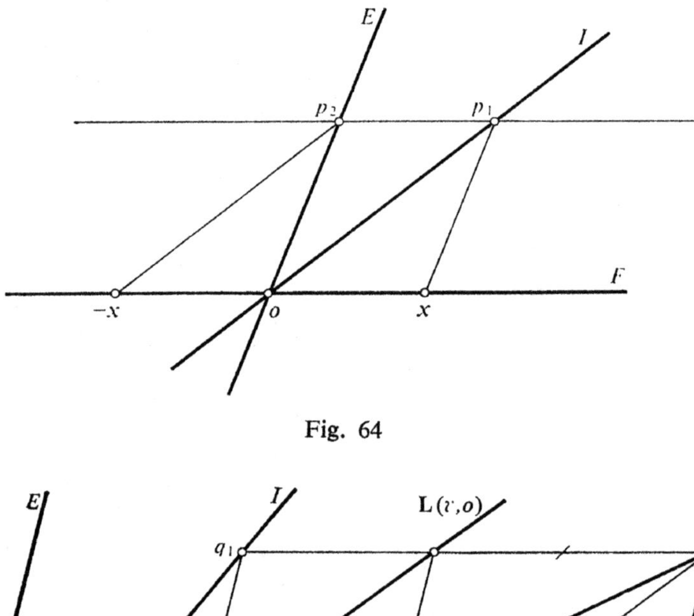

Fig. 64

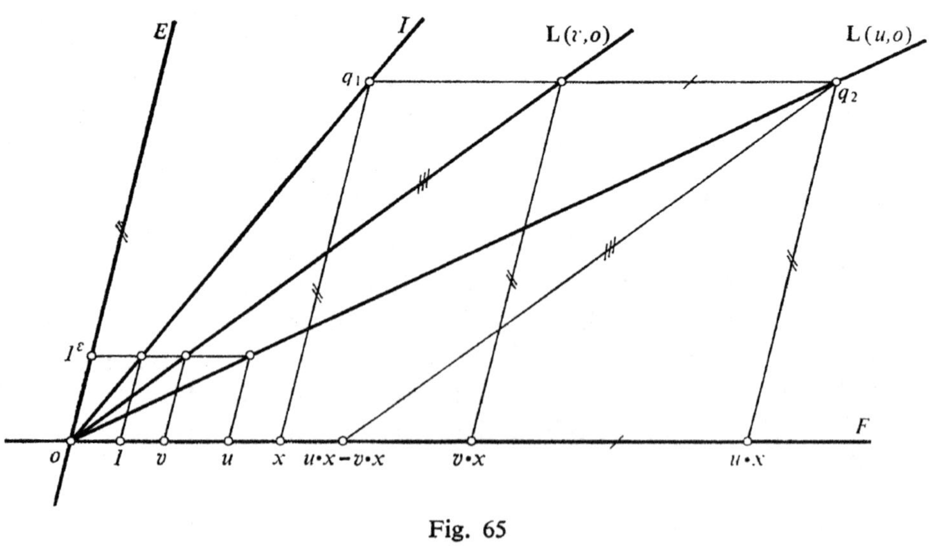

Fig. 65

Thus, by the law Pasch, all the functions σ^u and the function μ preserve the relation $\boldsymbol{B}|F$, and so they are monotone in each of two ordered sets $(F, \leqslant)$ and $(F, \geqslant)$. Hence, by 7.3.2, each of the relations $\leqslant$ and $\geqslant$ is an ordering of the group $\mathfrak{F}_+$, i.e. each of the structures $(\mathfrak{F}_+, \leqslant)$ and $(\mathfrak{F}_+, \geqslant)$ is an ordered group.

Let us pass to the quasi-field $\mathfrak{F}(o, 1, e)$. Given the points o and 1, we can distinguish from among $\leqslant$ and $\geqslant$ this relation to which the pair $(o, 1)$ belongs. Assume, this is the relation $\leqslant$; so

$$o \leqslant 1.$$

Let the functions $\varrho^{uv}\colon F \to F$ be defined by the formula

$$\varrho^{uv}(x) = u\cdot x - v\cdot x \quad \text{for} \quad u, v \in F.$$

Each of them, for $u \neq v$, is the composition of three parallel projections in $\mathfrak{A}$ (see fig. 65),

$$\varrho^{uv}\colon F \underset{||E}{\to} I \underset{||F}{\to} \mathbf{L}(u, o) \underset{\mathbf{L}||(v,o)}{\to} F$$
$$x \mapsto q_1 \mapsto q_2 \mapsto u\cdot x - v\cdot x.$$

Hence, by the law Pasch, ϱ^{uv} preserves the relation $\boldsymbol{B}|F$ and thus it is monotone in the ordered set $(F, \leqslant)$. Moreover, for $u \leqslant v$ the function ϱ^{uv} is decreasing; indeed, in this case $u - v \leqslant o$ and thus

$$o \leqslant 1 \quad \text{and} \quad u\cdot 1 - v\cdot 1 \leqslant u\cdot o - v\cdot o.$$

As a result, for arbitrary scalars $u, v, x, y \in F$

$$u \leqslant v \wedge x \leqslant y \Rightarrow u\cdot y - v\cdot y \leqslant u\cdot x - v\cdot x$$
$$\Rightarrow u\cdot y + v\cdot x \leqslant u\cdot x + v\cdot y,$$

whence, by 7.5.4, the relation $\leqslant$ is an ordering of the quasi-field $\mathfrak{F}(o, 1, e)$, i.e. $(\mathfrak{F}(o, 1, e), \leqslant)$ is an ordered quasi-field. Recall that the converse relation $\geqslant$ is not an ordering of the quasi-field $\mathfrak{F}(o, 1, e)$.

Since $\leqslant$ is an ordering of the group $\mathfrak{F}_+$, there are no elements of a finite order in the group Tr_F (which is isomorphic to $\mathfrak{F}_+$), as thus neither in the whole translation group Tr (comp. 7.4.1). Therefore, the plane $\mathfrak{A}$ has the characteristic $= 0$ (and so $\neq 2$) and thus it may be treated as a midpoint plane (see § 6.3). Hence, the ordered translation plane may be treated as an *ordered midpoint plane*, $(\boldsymbol{S}, \oplus, ||, \boldsymbol{B})$.[2] The

[2] The axiom 5.2.4 for midpoint algebras can be replaced by the simpler axiom 5.2.5; it follows from the fact that the group Tr has no elements of order 3 (see Hauschild and Rautenberg 74]).

axiom 8.2.0 is now superflous, since it follows from the axioms for midpoint planes. The axiom 7.1.10 is also superflous (comp. § 7.6). Hence, the set of sentences

$$\text{AX } 8.2 = \text{AX } 6.3 \cup \{7.1.8,\ 7.1.9\} \cup \{8.1.0,\ 8.1.1,\ \text{oPasch}\}$$

is an axiom system for the ordered midpoint planes.

Since the relation $\leqslant$ is an ordering of $\mathfrak{F}_+$, the relation $\boldsymbol{B}$ (more precisely, $\boldsymbol{B}|F$) is an ordering of the midpoint algebra $(F, \oplus)$ (comp. § 7.6). Because of free choice of the parameters o and 1 (and hence also of the line $F = [o1]$), for every line K on $\mathfrak{A}$ the relation $\boldsymbol{B}|K$ is an ordering of the midpoint algebra $(K, \oplus)$, i.e. the structure $(K, \oplus, \boldsymbol{B}|K)$ is an ordered midpoint algebra. We shall say less precisely that every line K on any ordered midpoint plane is an ordered midpoint algebra.

Hence, on the basis of the axioms for midpoint planes and betweenness relations, we have the following inference:

Th 8.2.1.[3] oPasch $\models\!\Rightarrow$ oInv.

When adding to AX 8.2 the axiom Des or Papp, we get an axiom system for *ordered Desarguean* or *Pappian midpoint planes*. We shall use the symbols

$$\text{OMid } \mathscr{A}, \quad \text{OMid Des } \mathscr{A}, \quad \text{OMid Papp } \mathscr{A}$$

for the class of the ordered midpoint planes, the ordered Desarguean midpoint planes, and the ordered Pappian midpoint planes, respectively. Clearly

$$\text{OMid } \mathscr{A} \supseteq \text{OMid Des } \mathscr{A} \supseteq \text{OMid Papp } \mathscr{A}$$

(comp. 6.3).

8.3. REPRESENTATION THEOREM

We are now going to connect the classes of planes OMid $\mathscr{A}$, OMid Des $\mathscr{A}$, and OMid Papp $\mathscr{A}$ with the classes of algebras **OQF**, **OF**, and com **OF** by means of the representation theorem.

Let us consider an arbitrary algebra $\mathfrak{F} = (F, 0, 1, +, \cdot) \in$ **QF** and an ordering $\leqslant$ of the set F. Let $\boldsymbol{B}$ be the betweenness relation on F,

[3] M.M.—to the end of § 8.3.

adjoint to this ordering. In the Cartesian structure over $\mathfrak{F}$

$$\underline{\mathfrak{A}}(\mathfrak{F}) = (F^2, \oplus_{\mathfrak{F}}, ||_{\mathfrak{F}}),$$

(see § 6.4) we define the relation $\boldsymbol{B}_{\mathfrak{F}}$ as follows: for any $a = (x_a, y_a)$, $b = (x_b, y_b)$, $c = (x_c, y_c) \in F^2$ we admit

Def 8.3.0. $\quad \boldsymbol{B}_{\mathfrak{F}}(abc) \Leftrightarrow \boldsymbol{L}_{\mathfrak{F}}(abc) \wedge \boldsymbol{B}(x_a x_b x_c) \wedge \boldsymbol{B}(y_a y_b y_c)$.[4]

We have the following

Theorem 8.3.1. *Let* $(\mathfrak{F}, \leqslant) \in \mathbf{OQF}$. *Then*

(i) $\quad \underline{\mathfrak{A}}(\mathfrak{F}) \in \mathrm{Mid}\, \mathscr{A}$,

(ii) $\quad \boldsymbol{B}_{\mathfrak{F}}$ *is a betweenness relation in* $\underline{\mathfrak{A}}(\mathfrak{F})$.

P r o o f. (i) follows from 6.4.2(i) combined with 7.5.8(i); (ii) follows by 8.3.0, 1.1.3′, 2.1.3, 4.0.2, 7.3.1, and 7.5.2. ∎

Theorem 8.3.2 (i) $\quad (\mathfrak{F}, \leqslant) \in \mathbf{OQF} \Rightarrow (\underline{\mathfrak{A}}(\mathfrak{F}), \boldsymbol{B}_{\mathfrak{F}}) \in \mathrm{OMid}\, \mathscr{A}$,

(ii) $\quad (\mathfrak{F}, \leqslant) \in \mathbf{OF} \Rightarrow (\underline{\mathfrak{A}}(\mathfrak{F}), \boldsymbol{B}_{\mathfrak{F}}) \in \mathrm{OMid\ Des}\, \mathscr{A}$,

(iii) $\quad (\mathfrak{F}, \leqslant) \in \mathrm{com}\, \mathbf{OF} \Rightarrow (\underline{\mathfrak{A}}(\mathfrak{F}), \boldsymbol{B}_{\mathfrak{F}}) \in \mathrm{OMid\ Papp}\, \mathscr{A}$.

P r o o f. (i) Let us first prove that the law oPasch is satisfied in $(\underline{\mathfrak{A}}(\mathfrak{F}), \boldsymbol{B}_{\mathfrak{F}})$.[5] Let

$$L = \{(x, y) \in F^2 : x = 0\}.$$

Take two lines K and K' with $K \cap K' = \{a\}$, and let

$$b, c \in K, \quad b'c' \in K', \quad bb' ||_{\mathfrak{F}} cc', \quad \text{and} \quad \boldsymbol{B}_{\mathfrak{F}}(abc).$$

If $[bb'] ||_{\mathfrak{F}} L$, then it is easy to check that $\boldsymbol{B}_{\mathfrak{F}}(ab'c')$. Assume $\sim([bb'] ||_{\mathfrak{F}} L)$; then, by 4.4.4, the equation of the line $[bb']$ has the form $y = u_0 \cdot x + v_0$. Since, by 7.5.4, the function $\varphi : F \to F$ defined by the formula

$$\varphi(x) = u \cdot x - u_0 \cdot x + v$$

is a monotone bijection for $u \neq u_0$ (comp. § 7.1), we easily prove that the parallel projections

$$\pi : K \underset{||[bb']}{\to} L \quad \text{and} \quad \pi' : L \underset{||[bb']}{\to} K'$$

preserve the relation $\boldsymbol{B}_{\mathfrak{F}}$; hence $\boldsymbol{B}_{\mathfrak{F}}(ab'c')$.

---

[4] To simplify the notation we use the symbol "$\boldsymbol{B}_{\mathfrak{F}}$" though this relation depends on the pair $(\mathfrak{F}, \leqslant)$ but not only on $\mathfrak{F}$.

[5] Compare Pickert [75], p. 236.

Therefore, by 8.3.1, $(\mathfrak{A}(\mathfrak{F}), \boldsymbol{B}_{\mathfrak{F}}) \in \text{OMid}\,\mathscr{A}$.

By (i), 6.4.2(ii) yields (ii), and 6.4.2(iii) yields (iii). ∎

If $(\mathfrak{F}, \leqslant) \in \mathbf{OQF}$, then the structure

$$\underline{\mathfrak{A}}(\mathfrak{F}, \leqslant) = (\underline{\mathfrak{A}}(\mathfrak{F}), \boldsymbol{B}_{\mathfrak{F}})$$

is referred to as the *Cartesian ordered midpoint plane over* $(\mathfrak{F}, \leqslant)$.

Consider now an arbitrary ordered midpoint plane $(\mathfrak{A}, \boldsymbol{B}) = (\boldsymbol{S}, \oplus, \|, \boldsymbol{B})$. Let $(\mathfrak{F}, \leqslant) = (\mathfrak{F}, o, 1, +, \cdot, \leqslant)$ be the ordered quasi-field with basis $(o, 1, e)$ (see § 8.2) and let $(\underline{\mathfrak{A}}(\mathfrak{F}), \boldsymbol{B}_{\mathfrak{F}})$ be the Cartesian ordered midpoint plane over $(\mathfrak{F}, \leqslant)$. We have the following obvious

Th 8.3.3.[6] $\boldsymbol{B}(abc) \Leftrightarrow \boldsymbol{B}_{\mathfrak{F}}(\Pi(a)\Pi(b)\Pi(c))$.
OMid

In view of 8.3.3, the argument given in § 8.2 leads to the following

Th 8.3.4. $(\mathfrak{A}, \boldsymbol{B}) \simeq (\underline{\mathfrak{A}}(\mathfrak{F}), \boldsymbol{B}_{\mathfrak{F}})$
OMid

and the function Π *is an isomorphism.*

Theorem 8.3.4 yields the representation theorem:

Theorem 8.3.5.

(i) $\text{OMid}\,\mathscr{A} \underset{\text{Iso}}{\subseteq} \{\underline{\mathfrak{A}}(\mathfrak{F}, \leqslant) : (\mathfrak{F}, \leqslant) \in \mathbf{OQF}\} \subseteq \text{OMid}\,\mathscr{A}$.

(ii) $\text{OMid Des}\,\mathscr{A} \underset{\text{Iso}}{\subseteq} \{\underline{\mathfrak{A}}(\mathfrak{F}, \leqslant) : (\mathfrak{F}, \leqslant) \in \mathbf{OF}\} \subseteq \text{OMid Des}\,\mathscr{A}$.

(iii) $\text{OMid Papp}\,\mathscr{A} \underset{\text{Iso}}{\subseteq} \{\underline{\mathfrak{A}}(\mathfrak{F}, \leqslant) : (\mathfrak{F}, \leqslant) \in \text{com}\,\mathbf{OF}\}$
$\subseteq \text{OMid Papp}\,\mathscr{A}$.

Proof. (i) by 8.3.2(i); (ii) by 4.3.9 and 8.3.2(ii); (iii) by 4.3.10 and 8.3.2(iii). ∎

Take $(\mathfrak{F}, \leqslant) \in \mathbf{OQF}$ and let $(\mathfrak{A}, \boldsymbol{B}) = \underline{\mathfrak{A}}(\mathfrak{F}, \leqslant)$. Let $F' = [(0, 0)\,(0, 1)]$ and let $(\mathfrak{F}', \leqslant')$ be the ordered quasi-field with basis $((0, 0), (0, 1), (1, 0))$. We shall prove

Th 8.3.6. $(\mathfrak{F}, \leqslant) \simeq (\mathfrak{F}', \leqslant')$.[7]

---

[6] The term $\underset{\text{OMid}}{\text{“Th”}}$ indicates that the theorem concerns the ordered midpoint planes.

[7] Compare 4.5.6.

Proof. By 4.5.6

$$\mathfrak{F} \simeq \mathfrak{F}',$$

and the function $f\colon F \to F'$ defined by the formula

$$f(x) = (0, x)$$

is an isomorphism.

The ordering $\leqslant$ generates the relation $B_{\mathfrak{F}}|[(0,0)(0,1)]$ on the set F',

(1) $\quad B_{\mathfrak{F}}((0,x)(0,y)(0,z)) \Leftrightarrow x \leqslant y \leqslant z \vee x \geqslant y \geqslant z$

(see 8.3.0). This relation is adjoint to the pair $(\leqslant', \geqslant')$, i.e.

(2) $\quad B_{\mathfrak{F}}((0,x)(0,y)(0,z)) \Leftrightarrow (0,x) \leqslant' (0,y) \leqslant' (0,z)$
$\qquad\qquad \vee (0,x) \geqslant' (0,y) \geqslant' (0,z).$

Moreover

(3) $\quad 0 < 1$

and

(4) $\quad (0,0) <' (0,1).$

Using (1)–(4) one can easily prove that

$$0 \leqslant x \Leftrightarrow (0,0) \leqslant' (0,x),$$

whence

(5) $\quad x \leqslant y \Leftrightarrow (0,x) \leqslant' (0,y).$

By (5), the bijection f is an isomorphism of the ordered quasi-fields Therefore

$$(\mathfrak{F}, \leqslant) \simeq (\mathfrak{F}', \leqslant'). \blacksquare$$

It was shown in § 8.2 that every ordered translation plane may be treated as an ordered midpoint plane. Thus the starting point was a structure $(\boldsymbol{S}, ||, \boldsymbol{B})$ satisfying the axiom oPasch. Next, we used this axiom to prove that the structure $(\mathfrak{F}(o, 1, e), \leqslant)$ is an ordered group. If, however, the operation $\oplus$ is given directly, i.e. we start with a midpoint plane $\mathfrak{A} = (\boldsymbol{S}, \oplus, ||)$ and a betweenness relation $\boldsymbol{B}$ in $(\boldsymbol{S}, ||)$ (comp. § 8.1), then we need only the law oInv which is weaker than oPasch (comp. 8.2.1).

Let us now take a triple of parameters, $(o, 1, e)$, and the line $F = [o1]$. Then $(F, \oplus)$ is a midpoint algebra (comp. § 6.3). If, moreover, the law oInv is satisfied in $(\mathfrak{A}, \boldsymbol{B})$, then the structure $(F, \oplus, \boldsymbol{B}|F)$ is an ordered midpoint algebra, and thus the algebra $(\mathfrak{F}_+(o, 1, e), \leqslant)$, adjoint to it, is an ordered group.

Hence, as a consequence of theorems 8.3.6, 8.3.4, and 8.3.2(i), we obtain

Theorem 8.3.7. *Let* $\mathfrak{A} = (\boldsymbol{S}, \oplus, \|)$ *be a midpoint plane and let* $\boldsymbol{B}$ *be a betweenness relation in* $(\boldsymbol{S}, \|)$. *If the law* oInv *is satisfied in* $(\mathfrak{A}, \boldsymbol{B})$, *then the following two conditions are equivalent:*

(i) *The law* oPasch *is satisfied in* $(\mathfrak{A}, \boldsymbol{B})$.

(ii) *The law of monotony of multiplication* (7.5.5) *is satisfied in the scalar algebra* $(\mathfrak{F}(o, 1, e), \leqslant)$ *of* $(\mathfrak{A}, \boldsymbol{B})$.[8]

All the above considerations deal with the affine planes described by elementary axioms. Adding to AX 8.2$\cup${Papp} the non-elementary axiom Arch′, i.e. assuming the relation $\boldsymbol{B}$ to be Archimedean in the midpoint algebra $(\boldsymbol{S}, \oplus)$, we obtain an axiom system for the ordered Cartesian midpoint plane over a commutative field with an Archimedean ordering, $(\mathfrak{F}, \leqslant)$ (see 8.3.5 (iii) and 7.6.5). It is well known[9] that such a field is isomorphic to a subfield of the ordered field of real numbers, $(\mathfrak{R}, \leqslant)$. Notice that adding the axiom Arch′ we can replace the axiom Papp by Des. Indeed, by 8.3.5(ii), any ordered Desarguean midpoint plane is (up to isomorphism) the structure $\underline{\mathfrak{A}}(\mathfrak{F}, \leqslant)$ over an ordered field $(\mathfrak{F}, \leqslant)$. By 7.6.5 combined with 8.3.3, since $\boldsymbol{B}$ is Archimedean in $(\boldsymbol{S}, \oplus)$, also $\leqslant$ is Archimedean in the group $\mathfrak{F}_+$, and thus, by 7.5.9, the field $\mathfrak{F}$ is commutative. In conclusion, the set of sentences

AX 8.2$\cup${Des, Arch′)}

is an axiom system for the Cartesian ordered midpoint planes over the subfields of $(\mathfrak{R}, \leqslant)$.

Adding to AX 8.2$\cup${Papp} the axiom Co′, i.e. assuming $\boldsymbol{B}$ to be continuous, we obtain an axiom system for the Cartesian ordered midpoint plane over the field $(\mathfrak{R}, \leqslant)$. Observe that we may now omit the axiom Papp. Indeed, by 8.3.5(i), an ordered midpoint plane is (up to

---

[8] Up to the present, the role of the Pasch axiom has been studied only for Euclidean geometry, but not, as here, for affine geometry (compare Szczerba [70]–[71d], Szczerba and Szmielew [70], Szmielew [70], [73], [74], [79]).

[9] Compare Van der Waerden [53], p.218, also Kurosh [73], p. 314.

isomorphism) the structure $\underline{\mathfrak{A}}(\mathfrak{F}, \leqslant)$ over a quasi-field $(\mathfrak{F}, \leqslant) \in \mathbf{OQF}$. By 7.1.21, since $\boldsymbol{B}$ is continuous, also the order $\leqslant$ is continuous in F, whence, by 7.5.11, $\mathfrak{F} \in \text{com } \mathbf{F}$. In conclusion, the set AX $8.2 \cup \{\text{Co}'\}$ is an axiom system for the plane $\underline{\mathfrak{A}}(\mathfrak{R}, \leqslant)$.

PART III

CHAPTER 9

EUCLIDEAN PLANES.

9.1. DESARGUEAN EUCLIDEAN PLANES

To pass from affine to Euclidean planes, we have to extend the system of affine primitive notions by a new, Euclidean, primitive notion. Usually the *equidistance* relation $\equiv$ (see fig. 66) is taken as such a notion. This relation is a basic invariant of similarities. In our approach equidistance is a quaternary relation on the set $\boldsymbol{S}$; the formula

$$ab \equiv cd$$

reads: *the point a is as distant from b as c is from d.*

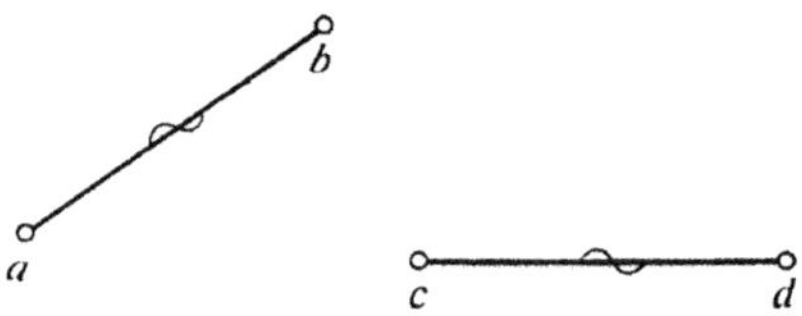

Fig. 66

Instead of equidistance, the *perpendicularity* relation $\perp$ (i.e. the relation of *orthogonal basis,* see fig. 67) may be added to $\mathfrak{A}$; the formula

$$\perp(apq)$$

reads: *the points a, p, q form a triangle with the right angle by the vertex a.*

Consider now the relation of *equi-orthogonal basis* (or, briefly, the relation of *basis*), $\boldsymbol{O}$ (see fig. 68); the formula

$$\boldsymbol{O}(apq)$$

reads: *the points a, p, q form a rectangle isosceles triangle with the vertex a*, or, *the configuration apq is a (equi-orthogonal) basis with the vertex a.*

Thus, in the notion of equi-orthogonal basis both relations, equidistance and perpendicularity, are involved.[1] However, this notion turns out to be fundamental for the coordinatization. Indeed, with the help of equi-orthogonal basis one obtains at once a rectangular coordinate system with congruent units on both axis; moreover, in such a coordinate system the basis relation can be directly described analytically. If, otherwise, either equidistance or perpendicularity is chosen as a primitive notion, then anyway one has to use the equi-orthogonal basis for an analytic description of the primitive notion.

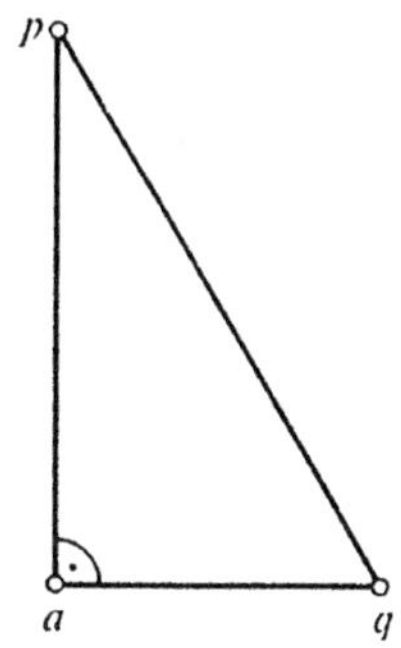

Fig. 67

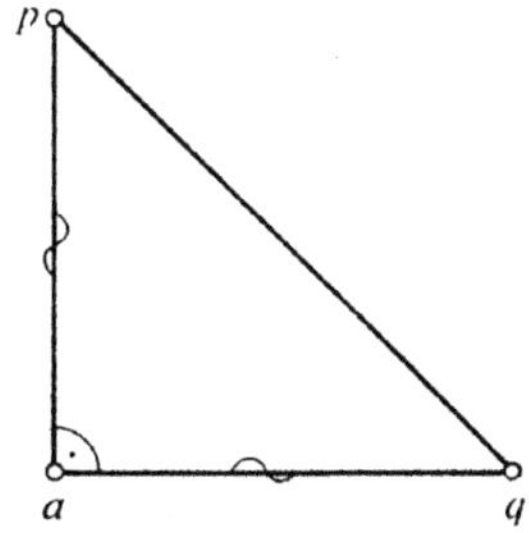

Fig. 68

The notion of equi-orthogonal basis is very weak, since, as is easy to foresee, in a rectangular coordinate system it can be described in terms of the additive group (of the field of scalars) itself. Thus, the connections between the relation $\boldsymbol{O}$ and the operation $\oplus$ can be also

[1] More precisely, the relation $\boldsymbol{O}$ is the common part of the perpendicularity relation, $\perp$, and the Pieri relation (i.e. the relation of isosceles triangle), $\triangle$, expressed in terms of equidistance by the formula: $\triangle(abc) \Leftrightarrow ab \equiv ac$.

expressed in terms of this group. Though to express the connections of $\boldsymbol{O}$ with $\|$ one needs already the whole scalar field, however, as will be shown, still the commutativity of multiplication is not needed. In conclusion, adding the relation $\boldsymbol{O}$ to a Desarguean midpoint plane, we obtain an Euclidean plane which does not need to satisfy the Pappus axiom. Let us recall that any Euclidean plane with equidistance or perpendicularity has to be Pappian.

For all the above reasons, we choose the relation $\boldsymbol{O}$ as the new Euclidean primitive notion. We now give the axioms for the relation $\boldsymbol{O}$

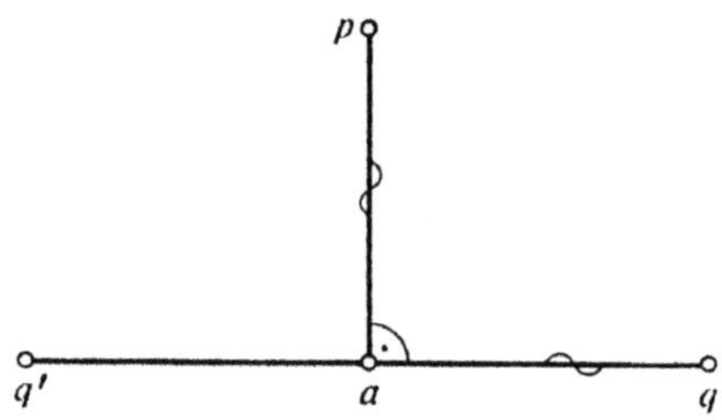

Fig. 69

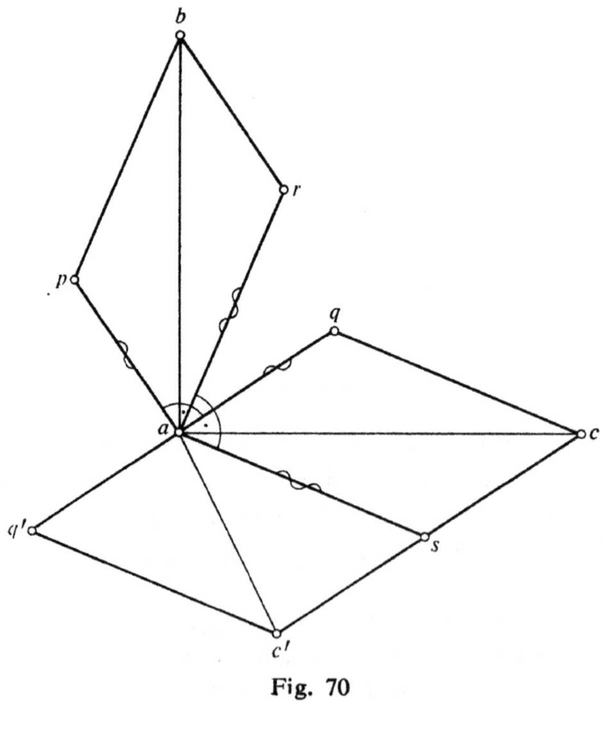

Fig. 70

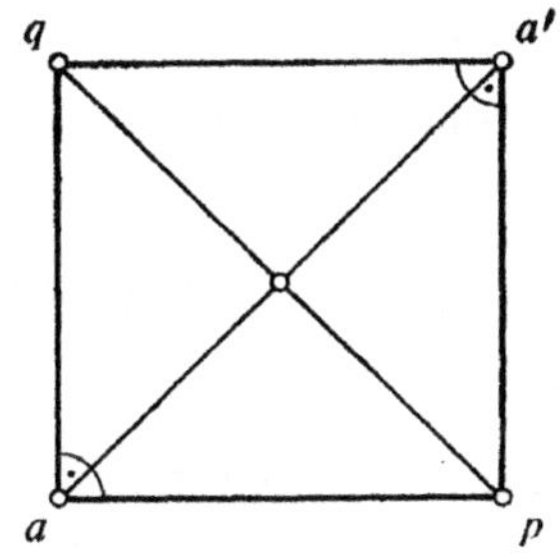

Fig. 71

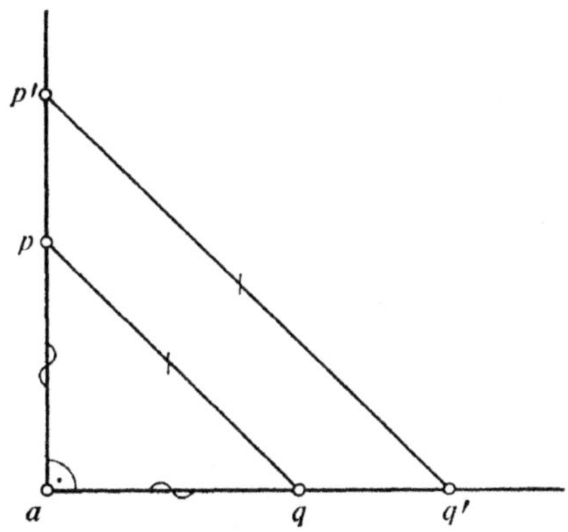

Fig. 72

and thereby define the class of Euclidean planes; then, in this section and the next two, we shall justify our intuitive supposition by precise proofs.

Let us consider a Desarguean midpoint plane $\mathfrak{A} = (\boldsymbol{S}, \oplus, \|)$ and a relation $\boldsymbol{O} \subseteq \boldsymbol{S} \times \boldsymbol{S} \times \boldsymbol{S}$. The expansion $\mathfrak{E} = (\boldsymbol{S}, \oplus, \|, \boldsymbol{O})$ of $\mathfrak{A}$ is a *Desarguean Euclidean plane* if and only if it satisfies the following axioms:

Ax 9.1.1. $\forall ap \, \exists q \; \boldsymbol{O}(apq)$,

Ax 9.1.2. $\boldsymbol{O}(apq) \Rightarrow \boldsymbol{O}(aqp)$,

Ax 9.1.3. $\boldsymbol{O}(apq) \wedge \boldsymbol{O}(apq') \Rightarrow q = q' \vee a = q \oplus q'$,

Ax 9.1.4. $\boldsymbol{O}(apq) \wedge a = q \oplus q' \Rightarrow \boldsymbol{O}(apq')$,

Ax 9.1.5. $\boldsymbol{O}(apq) \wedge a = q \oplus q' \wedge \boldsymbol{O}(asr) \wedge p \oplus r = a \oplus b$
$\wedge q \oplus s = a \oplus c \wedge q' \oplus s = a \oplus c' \Rightarrow \boldsymbol{O}(abc) \vee \boldsymbol{O}(abc')$,

Ax 9.1.6. $O(apq) \wedge O(a'pq) \Rightarrow a = a' \vee a \oplus a' = p \oplus q$,

Ax 9.1.7. $a \neq p \wedge O(apq) \wedge L(app') \wedge L(aqq') \wedge pq \| p'q'$
$\Rightarrow O(ap'q')$,

Ax 9.1.8. $a \neq p \wedge O(apq) \wedge L(paa') \wedge L(pqq') \wedge aq \| a'q'$
$\Rightarrow O(a'pq')$.

The figure 69 illustrates the axioms 9.1.1–9.1.4, while the figures 70, 71, 72, and 73 illustrate the axioms 9.1.5, 9.1.6, 9.1.7, and 9.1.8, respectively. The axiom 9.1.5 reads briefly as follows: *the sum of two basis with the same orientation and the common vertex* (*apq* and *ars* on fig. 70) *is again a basis with the same vertex* (*abc* on fig. 70).

The following three theorems deal with degenerate basis.

Th E 9.1.9.[2] $O(aaa)$.
Mid Des

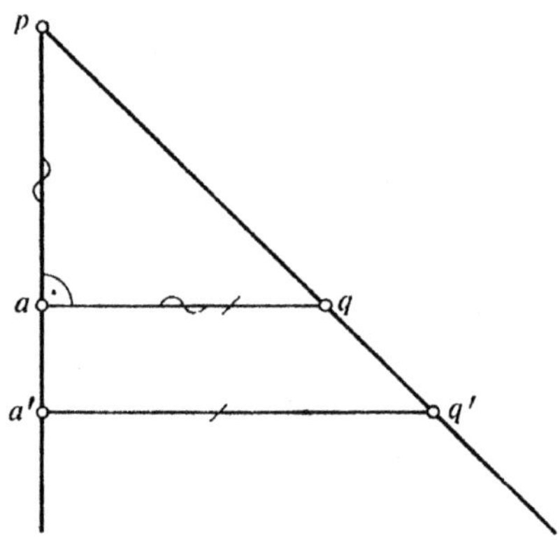

Fig. 73

Proof. Take an arbitrary point $p \neq a$. By the axiom 9.1.1, there exists a point q such that $O(apq)$. By 9.1.7

$$a \neq p \wedge O(apq) \wedge L(apa) \wedge L(aqa) \wedge pq \| aa \mapsto O(aaa). \blacksquare$$

Th E 9.1.10. $O(aaq) \Rightarrow a = q$.
Mid Des

Proof. By 9.1.9, $O(aaa)$. By 9.1.3

$$O(aaa) \wedge O(aaq) \mapsto a = q \vee a = a \oplus q \mapsto a = q. \blacksquare$$

[2] The term "Th E" / Mid Des indicates Desarguean Euclidean planes, to which the theorem refers (M.M.).

Th E 9.1.11. $O(apq) \wedge L(apq) \Rightarrow a = p$.
Mid Des

Proof. By 9.1.7 and 9.1.10

$$a \neq p \wedge O(apq) \wedge L(apa) \wedge L(aqp) \wedge pq||ap \mapsto O(aap) \mapsto a = p. \blacksquare$$

Hence, a degenerate basis is always reduced to a point, while a non-degenerate basis always consists of three non-collinear points.

We are now going to construct a Cartesian Desarguean Euclidean plane. Take an arbitrary (non-necessarily commutative) field $\mathfrak{F} = (F, 0, 1, +, \cdot)$ in which -1 is not a square, i.e.

(i) $\quad 1+u^2 \neq 0 \quad$ for $\quad u \in F$.

The condition (i) is equivalent to

(ii) $\quad u_1^2+u_2^2 = 0 \Rightarrow u_1 = u_2 = 0$.

A field $\mathfrak{F}$ satisfying this condition is said to be *2-formally real*;[3] it has, of course, the characteristic $\neq 2$, because $1+1 = 0$ contradicts the condition (i).

The field $\mathfrak{F}$ generates the Cartesian Desarguean midpoint plane

$$\underline{\mathfrak{A}}(\mathfrak{F}) = (F^2, \oplus_{\mathfrak{F}}, ||_{\mathfrak{F}})$$

(comp. 6.4.2(ii)). On the universe F^2 of this structure we now construct a new ternary relation $O_{\mathfrak{F}}$. Let $a, p, q \in F^2$ and let $a = (a_1, a_2)$, $p = (p_1, p_2)$, $q = (q_1, q_2)$; then

Def 9.1.12. $O_{\mathfrak{F}}(apq) \Leftrightarrow$

$$(p_1-a_1) = (q_2-a_2) \wedge (p_2-a_2) = -(q_1-a_1)$$
$$\vee (p_1-a_1) = -(q_2-a_2) \wedge (p_2-a_2) = (q_1-a_1).$$

Adding to $\underline{\mathfrak{A}}(\mathfrak{F})$ the relation $O_{\mathfrak{F}}$, we obtain the new Cartesian structure

$$\underline{\mathfrak{E}}(\mathfrak{F}) = (F^2, \oplus_{\mathfrak{F}}, ||\mathfrak{F}, O_{\mathfrak{F}})$$

generated by $\mathfrak{F}$. We shall show that $\underline{\mathfrak{E}}(\mathfrak{F})$ is a Desarguean Euclidean plane, i.e. in addition to the axioms for Desarguean midpoint planes it satisfies all the axioms of equi-orthogonal basis. We leave to the

[3] Generally, a field $\mathfrak{F}$ is said to be *k-formally real* whenever $u_1^2+ \ldots +u_k^2 = 0 \Rightarrow$ $\Rightarrow u_1 = \ldots = u_k = 0$. In the literature, any field which is k-formally real for all k is referred to as a *formally real field* (compare e.g. Van der Waerden [53], p. 225) (M.M).

reader the axioms 9.1.1–9.1.6, which can be verified by a simple calculation in the additive group $\mathfrak{F}_+$ of the field $\mathfrak{F}$. We are going to verify the axioms 9.1.7 and 9.1.8. First of all let us show that

(0) $\quad \boldsymbol{O}_{\mathfrak{F}}(apq) \wedge \boldsymbol{L}_{\mathfrak{F}}(apq) \Rightarrow a = p$

(comp. 9.1.11). Indeed, let $a, p, q \in F^2$. Assume $\boldsymbol{O}_{\mathfrak{F}}(apq)$ and $\boldsymbol{L}_{\mathfrak{F}}(apq)$ and suppose $a \neq p$. Then

$$(p_1 - a_1) = (q_2 - a_2) \quad \text{and} \quad (p_2 - a_2) = -(q_1 - a_1)$$

(1) or

$$(p_1 - a_1) = -(q_2 - a_2) \quad \text{and} \quad (p_2 - a_2) = (q_1 - a_1)$$

and, for some $u \in F$,

$$(p_1 - a_1) \cdot u = (q_1 - a_1) \quad \text{and} \quad (p_2 - a_2) \cdot u = (q_2 - a_2).$$

Thus

$$(p_1 - a_1) = (p_2 - a_2) \cdot u \quad \text{and} \quad (p_2 - a_2) = -(p_1 - a_1) \cdot u$$

or

$$(p_1 - a_1) = -(p_2 - a_2) \cdot u \quad \text{and} \quad (p_2 - a_2) = (p_1 - a_1) \cdot u;$$

hence

$$(p_1 - a_1) \cdot (1 + u^2) = 0 \quad \text{and} \quad (p_2 - a_2) \cdot (1 + u^2) = 0.$$

Since $\mathfrak{F}$ is 2-formally real, we have $1 + u^2 \neq 0$, and therefore $a = p$.

Let us pass to the axiom 9.1.7. Let $a, p, q, p', q' \in F^2$. Assume $a \neq p$, $\boldsymbol{O}_{\mathfrak{F}}(apq)$, $\boldsymbol{L}_{\mathfrak{F}}(app')$, $\boldsymbol{L}_{\mathfrak{F}}(aqq')$, and $pq \|_{\mathfrak{F}} p'q'$. Since the structure $\underline{\mathfrak{E}}(\mathfrak{F})$ satisfies the condition (0), we have

(2) $\quad \sim \boldsymbol{L}_{\mathfrak{F}}(apq).$

Moreover, the condition (1) and, for some $u, v, t \in F$, the conditions

(3) $\quad (p' - a) = (p - a) \cdot u,$

(4) $\quad (q' - a) = (q - a) \cdot v,$

and

(5) $\quad (p' - q') = (p - q) \cdot t$

hold. By (5), we get

$$(p' - a) - (q' - a) = ((p - a) - (q - a)) \cdot t,$$

which, combined with (3) and (4), implies

$$(p-a)\cdot u-(q-a)\cdot v = (p-a)\cdot t-(q-a)\cdot t$$

and

$$(p-a)\cdot (u-t) = (q-a)\cdot (v-t).$$

Thus, by (2), we obtain

$$u = v = t,$$

which, combined with (1), (3), and (4), yields

$$(p_1'-a_1) = (q_2'-a_2) \quad \text{and} \quad (p_2'-a_2) = -(q_1'-a_1)$$

or

$$(p_1'-a_1) = -(q_2'-a_2) \quad \text{and} \quad (p_2'-a_2) = (q_1'-a_1),$$

i.e. $O_{\mathfrak{F}}(ap'q')$. Hence the axiom 9.1.7 is satisfied in $\underline{\mathfrak{E}}(\mathfrak{F})$.

By the similar argument, the axiom 9.1.8 is also satisfied. In conclusion, all the axioms of equi-orthogonal basis are satisfied in $\underline{\mathfrak{E}}(\mathfrak{F})$. Therefore, we have

Theorem 9.1.13. *If the field $\mathfrak{F}$ is 2-formally real, then the structure $\underline{\mathfrak{E}}(\mathfrak{F})$ is a (Cartesian) Desarguean Euclidean plane.*

9.2. EQUI-ORTHOGONAL BASIS

We shall now establish the properties of basis which are essential for coordinatization of Euclidean planes (see § 9.3).

The axioms 9.1.7 and 9.1.8 are particular cases of the general law 9.2.3 (see p. 180). The following two theorems (deduced from 9.1.7 and 9.1.8 by means of 9.1.1–9.1.4) are also its particular cases.

Th E 9.2.1. $O(apq) \wedge p\oplus q = a\oplus a' \Rightarrow O(a'qp)$.
Mid Des

P r o o f. Assume

$$O(apq) \quad \text{and} \quad p\oplus q = a\oplus a'.$$

If $a = p$, then, by 9.1.10, $a = p = q$ and thus $a' = p = q$, whence, by 9.1.9, $O(a'pq)$. If $a \neq p$ (see fig. 74), then, by the axiom 9.1.8 combined with the theorem 6.2.12,

$$O(apq) \mapsto O(p\oplus a\ p\ p\oplus q) \mapsto O(a'\oplus q\ p\ p\oplus q) \mapsto O(a'qp). \blacksquare$$

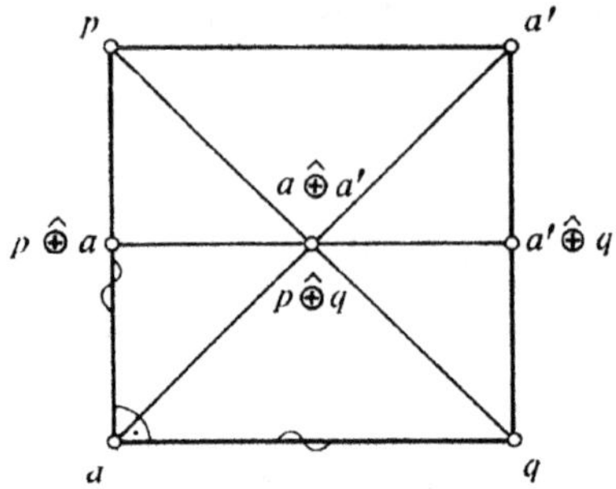

Fig. 74

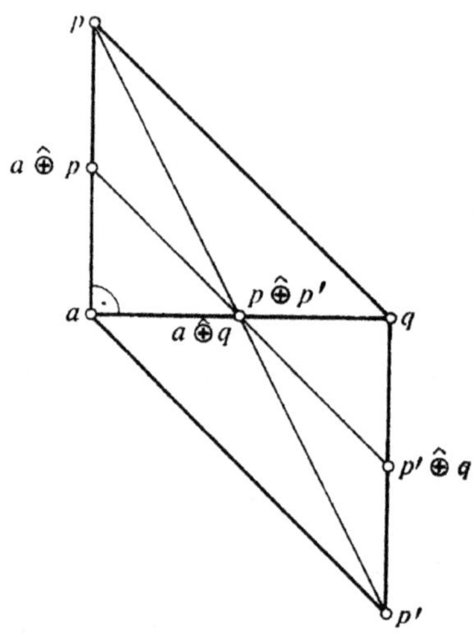

Fig. 75

Th E 9.2.2. $O(apq) \wedge a \oplus q = p \oplus p' \Rightarrow O(qp'a)$.
Mid Des

P r o o f. For $a = p$ this is a direct consequence of 9.1.9 combined with 9.1.10. Assume

$$a \neq p, \quad O(apq) \quad \text{and} \quad a \oplus q = p \oplus p'$$

(see fig. 75). Then, by 9.1.11, $a \neq q$ and thus, by 9.1.7 and 9.1.8,

$$O(apq) \mapsto O(a\ a \oplus p\ a \oplus q) \mapsto O(q\ a \oplus q\ p' \oplus q) \mapsto O(qap'). \blacksquare$$

We shall now establish a general theorem, which can be reduced (by means of the axioms 9.1.1–9.1.4) to its four particular cases 9.1.7,

9.1.8, 9.2.1, and 9.2.2, and thus, indirectly, to the axioms 9.1.7 and 9.1.8 themselves.

Th E 9.2.3. $a \neq p \wedge \boldsymbol{O}(apq) \wedge ap||a'p' \wedge aq||a'q' \wedge pq||p'q' \Rightarrow \boldsymbol{O}(a'p'q')$. **Mid Des**

Proof. Assume

$$a \neq p, \quad \boldsymbol{O}(apq), \quad ap||a'p', \quad aq||a'q', \quad \text{and} \quad pq||p'q'.$$

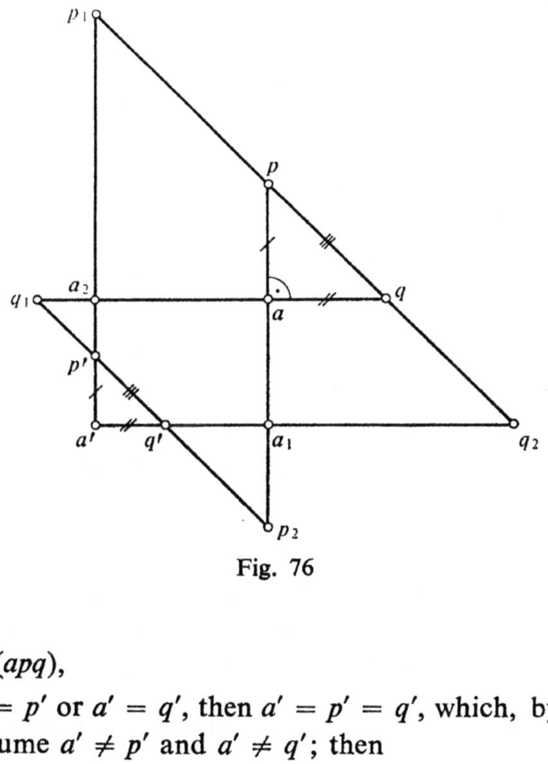

Fig. 76

By 9.1.11,

$$\sim \boldsymbol{L}(apq),$$

whence, if $a' = p'$ or $a' = q'$, then $a' = p' = q'$, which, by 9.1.9, yields $\boldsymbol{O}(a'p'q')$. Assume $a' \neq p'$ and $a' \neq q'$; then

$$\sim \boldsymbol{L}(a'p'q').$$

Let (see fig. 76)

$$[ap] \cap [a'q'] = \{a_1\},$$
$$[ap] \cap [p'q'] = \{p_2\},$$
$$[aq] \cap [p'q'] = \{q_1\},$$
$$[aq] \cap [a'p'] = \{a_2\},$$
$$[pq] \cap [a'p'] = \{p_1\},$$
$$[pq] \cap [a'q'] = \{q_2\}.$$

If

(1) $\quad p \notin [a'q'] \quad$ and $\quad a' \notin [pq]$,

then $a_1 \neq p$ and $a' \neq p_1$, and therefore on the basis of 9.1.8 and 9.1.7

$$O(apq) \mapsto O(a_1pq_2) \mapsto O(a'p_1q_2) \mapsto O(a'p'q').$$

If

(2) $\quad q \notin [a'p'] \quad$ and $\quad a' \notin [pq]$,

or

(3) $\quad p' \notin [aq] \quad$ and $\quad a \notin [p'q']$,

or

(4) $\quad q' \notin [ap] \quad$ and $\quad a \notin [p'q']$,

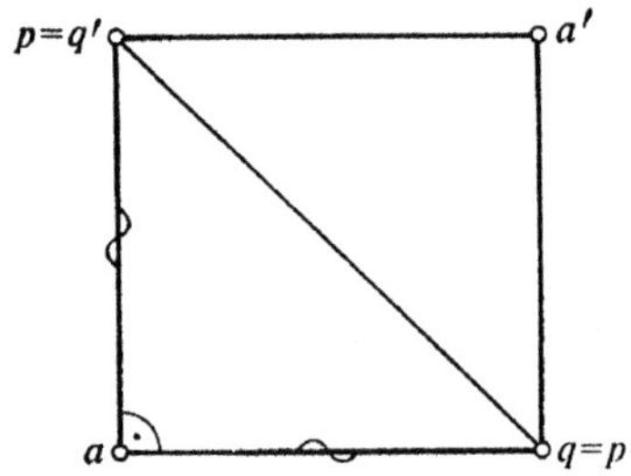

Fig. 77

then, by the similar argument, $O(apq)$ implies $O(a'p'q')$. Assume finally that no one of the conditions (1)–(4) holds. This case can be easily reduced to the following four particular cases, (i)–(iv).

(i) $\quad p \in [a'q'], \quad q \in [a'p'], \quad p' \in [aq], \quad q' \in [ap].$

Then $a_1 = p = q'$, $a_2 = q = p'$, $p \oplus q = a \oplus a'$ (see fig. 77), and thus, by 9.2.1,

$$O(apq) \mapsto O(a'qp) \mapsto O(a'p'q').$$

(ii) $\quad a' \in [pq] \quad$ and $\quad a \in [p'q']$.

Then $a = q'$ implies $q = a'$ and $a \oplus q = p \oplus p'$ (see fig. 78), whence by 9.2.2 (see fig. 79)

$$O(apq) \mapsto O(qp'a) \mapsto O(a'p'q');$$

if $a = p'$, then the similar reasoning leads to the same conclusion; if

$a \neq p', q'$, then $a_1 \neq p, a$ and, by 9.1.8 and 9.1.7,

$$O(apq) \mapsto O(a_1 pa') \mapsto O(a_1 aq') \mapsto O(a'p'q').$$

(iii) $\quad p \in [a' q'], \quad q \in [a'p'], \quad a \in [p'q'].$

Then (see fig. 80) $p \oplus q = a \oplus a'$ and $a' \neq q$, whence by 9.2.1 combined with 9.1.7

$$O(apq) \mapsto O(a'qp) \mapsto O(a'p'q').$$

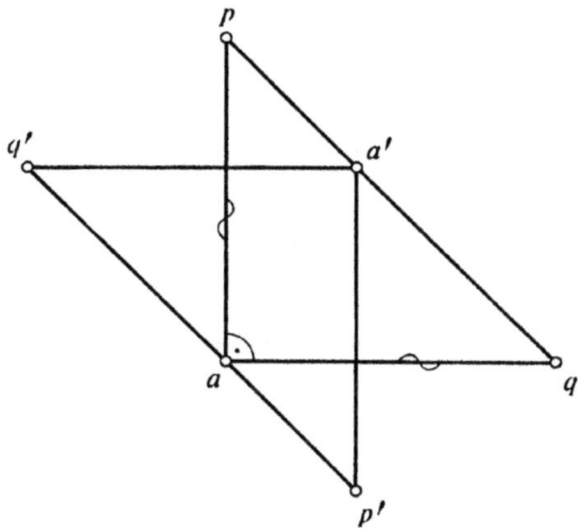

Fig. 78

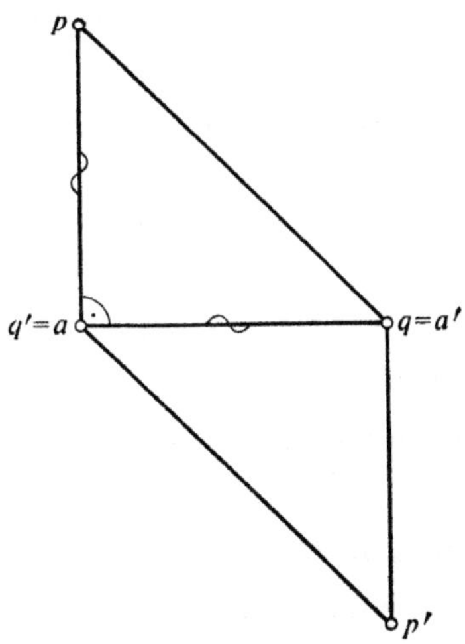

Fig. 79

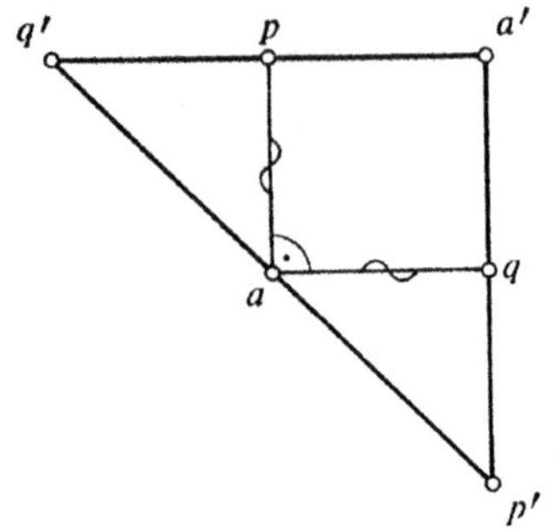

Fig. 80

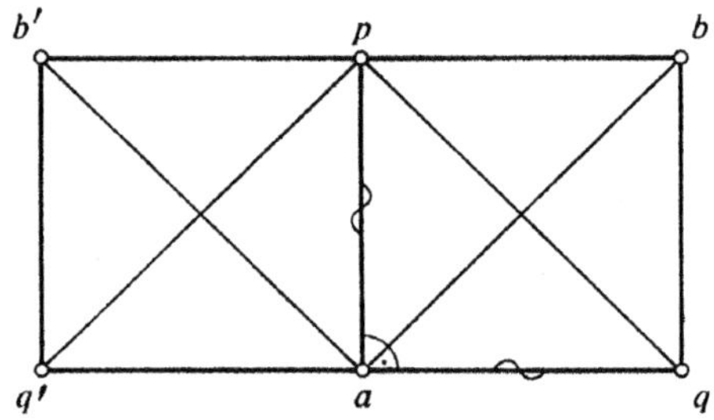

Fig. 81

Also in the last case,

(iv) $p' \in [aq], \quad q' \in [ap], \quad a' \in [pq],$

by the similar argument we obtain $\boldsymbol{O}(a'p'q')$. Thus, in any case $\boldsymbol{O}(a'p'q')$. ∎

From 9.2.3 it follows immediately that the notion of basis is an invariant of the dilatation group and thus, in particular, of any reflection (comp. § 6.3):

Th E 9.2.4. $\boldsymbol{O}(apq) \wedge a \oplus a' = p \oplus p' = q \oplus q' \Rightarrow \boldsymbol{O}(a'p'q')$. ∎
Mid Des

Till now we have not used the axioms 9.1.5 and 9.1.6. The next two theorems are based on 9.1.5.

Th E 9.2.5. $\boldsymbol{O}(apq) \wedge a = q \oplus q' \wedge p \oplus q = a \oplus b \wedge p \oplus q' = a \oplus b'$
Mid Des $\Rightarrow \boldsymbol{O}(abb')$.

Proof. Let all the assumptions be satisfied (see fig. 81). Then $\boldsymbol{O}(aq'p)$ and, by 9.1.5,

$$(\boldsymbol{O}(apq) \wedge a = q \oplus q' \wedge \boldsymbol{O}(aq'p)$$
$$\wedge p \oplus q' = a \oplus b' \wedge q \oplus p = a \oplus b \wedge q' \oplus p = a \oplus b')$$
$$\mapsto \boldsymbol{O}(ab'b) \vee \boldsymbol{O}(ab'b') \mapsto \boldsymbol{O}(abb') \vee a = b' = b \mapsto \boldsymbol{O}(abb'). \blacksquare$$

Moreover, by a simple calculation in the midpoint algebra we obtain the formula

$$a \oplus p = b' \oplus q = b \oplus q',$$

which, by 9.2.4, enables us to pass from the basis abb' to the new basis pqq'. Thus, as a direct consequence of 9.2.5 we get

Th E 9.2.6. $\boldsymbol{O}(apq) \wedge a = q \oplus q' \Rightarrow \boldsymbol{O}(pqq')$.
Mid Des

9.3. COORDINATIZATION

Let $\mathfrak{E} = (\boldsymbol{S}, \oplus, \|, \boldsymbol{O})$ be a Desarguean Euclidean plane. Then the structure $\mathfrak{A} = (\boldsymbol{S}, \oplus, \|)$ is a Desarguean midpoint plane, and two parameters o, 1 determine the scalar field $\mathfrak{F} = (F, o, 1, +, \cdot)$ (commutative or not) which is independent of the additional parameter e. Assume

(i) $\boldsymbol{O}(o1e)$;

then $\sim \boldsymbol{L}(o1e)$. Let $a = \mathbf{P}(o1e)$ and let

$$E = [oe] \quad \text{and} \quad I = [oa]$$

(see def 4.1.1). By 6.3.2 and 6.3.4

$$o \oplus a \in I \quad \text{and} \quad 1 \oplus e = o \oplus a$$

(see fig. 82), whence

(ii) $1 \oplus e \in I$.

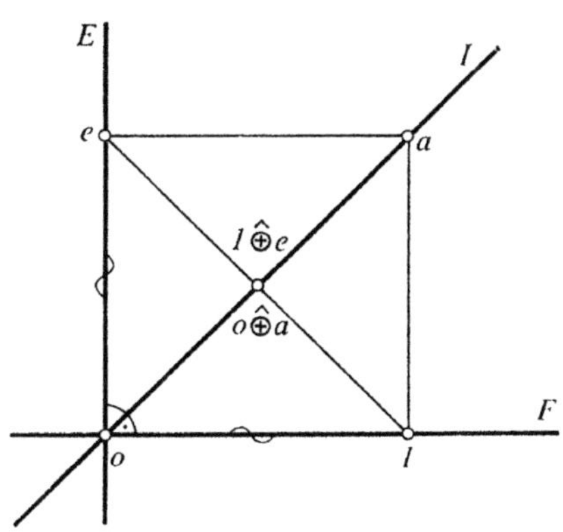

Fig. 82

According to (i) and (ii), the Cartesian coordinate system Π with basis $(o, 1, e)$ will be referred to as a rectangular coordinate system. We preserve the notation introduced before for arbitrary Cartesian coordinate system:

$$\Pi(p) = (x_p, y_y), \qquad \Pi(a) = (1, 1),$$

$$p = p(x_p, y_p), \qquad a = p(1, 1).$$

Generally, for $x \in F$ let (see fig. 83)

$$a^x = p(x, x)$$

(in particular $a = a^1$); then, by 4.2.7(iii),

$$a^x \in I.$$

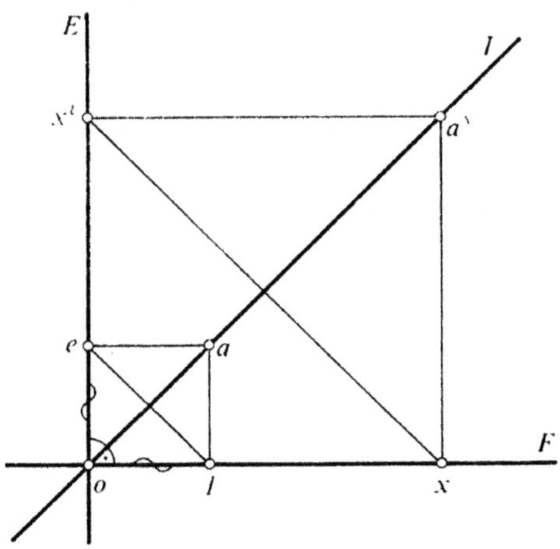

Fig. 83

Applying the law Des′ to the trapezoids aa^x1x and aa^xex^e, we obtain

(iii) $\quad 1e \| xx^e.$

By 9.1.7, the formulae (i) and (iii) yield

(iv) $\quad O(oxx^e),$

and thus, by 9.1.2, also

(iv′) $\quad O(ox^ex).$

Take now an arbitrary point $p \in S$. Evidently (see fig. 84) the point x_p^ε is the image of p under the projection parallel to F. Therefore

(v) $\qquad x_p^\varepsilon \oplus y_p = o \oplus p.$

In a rectangular coordinate system (as in every Cartesian system) the operation $\oplus$ and the relation $\|$ can be expressed analytically by the formulae 6.3.8 and 4.4.6, respectively. We now pass to the description of the relation O (by use of 9.1.6).

Th E 9.3.1. $\quad O(opq) \Rightarrow x_p = y_q \wedge y_p = -x_q \vee x_p = -y_q \wedge y_p = x_q.$
Mid Des

Proof. Take $p, q \in S$. By (iv), (iv′), and (v)

$$O(ox_p^\varepsilon x_p), \quad O(oy_p y_p^\varepsilon) \quad \text{and} \quad x_p^\varepsilon \oplus y_p = o \oplus p.$$

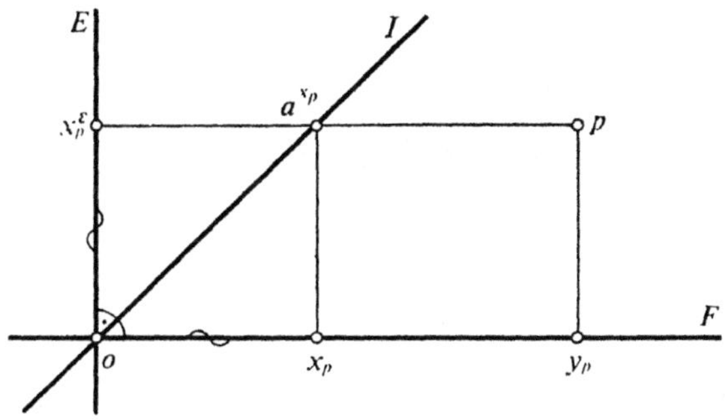

Fig. 84

Let (see fig. 85)

$$x_p \oplus y_p^\varepsilon = o \oplus r \quad \text{and} \quad (-x_p) \oplus y_p^\varepsilon = o \oplus s;$$

then, by 9.1.5, we obtain

(1) $\qquad O(opr) \quad \text{or} \quad O(ops).$

Obviously

$$\Pi(r) = (y_p, x_p) \quad \text{and} \quad \Pi(s) = (y_p, -x_p);$$

let

$$\Pi(r') = (-y_p, -x_p) \quad \text{and} \quad \Pi(s') = (-y_p, x_p),$$

and let (see fig. 85)

$$\Pi(c) = (y_p, y_p) \quad \text{and} \quad \Pi(c') = (-y_p, y_p).$$

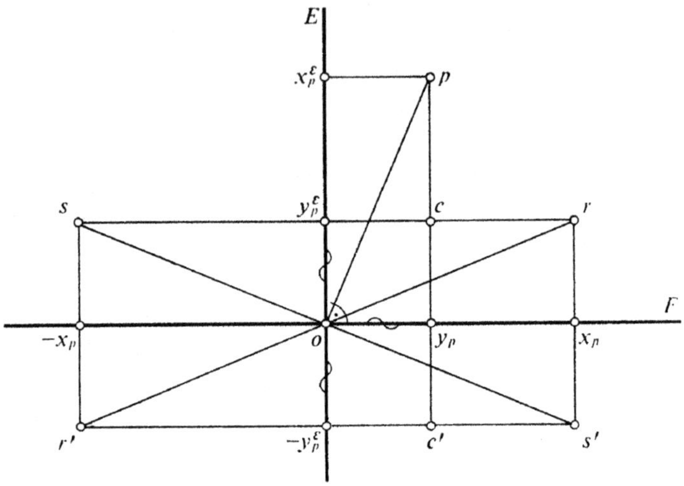

Fig. 85

Then, by 9.1.4 and 9.2.3,

$$\boldsymbol{O}(cy_py_p^\varepsilon) \quad \text{and} \quad \boldsymbol{O}(c'\,y_p\,-y_p^\varepsilon),$$

whence, by 9.1.7, we easily get

(2) $\boldsymbol{O}(cpr) \wedge \boldsymbol{O}(c'pr')$.

By 9.1.6 and (2)

$$\boldsymbol{O}(opr) \Rightarrow \boldsymbol{O}(opr) \wedge \boldsymbol{O}(cpr) \Rightarrow o = c \vee o \oplus c = p \oplus r$$
$$\Rightarrow y_p = o \vee x_p = o$$

and

$$\boldsymbol{O}(opr') \Rightarrow \boldsymbol{O}(opr') \wedge \boldsymbol{O}(c'pr') \Rightarrow o = c' \vee o \oplus c' = p \oplus r'$$
$$\Rightarrow y_p = o \vee x_p = o.$$

Assume now

(3) $\boldsymbol{O}(opq)$.

From (1) and (3), by the axiom 9.1.3, it follows at once that the point q is one of the points r, r', s, s'. If $q = s$ or $q = s'$, then

(4) $x_p = y_q \wedge y_p = -x_q \vee x_p = -y_q \wedge y_p = x_q$;

if $q = r$ or $q = r'$, then

$$x_p = y_q \wedge y_p = x_q \vee x_p = -y_q \wedge y_p = -x_q$$

and simultaneously $x_q = o$ or $y_q = o$. Thus in any case the condition (4) is satisfied. ∎

Th E 9.3.2. $x_p = y_q \wedge y_p = -x_q \vee x_p = -y_q \wedge y_p = x_q \Rightarrow \boldsymbol{O}(opq).$
Mid Des

P r o o f. Assume

(5.1) $\quad x_p = y_q \quad$ and $\quad y_p = -x_q.$

By the axiom 9.1.1, for some point r

(6) $\quad \boldsymbol{O}(opr).$

Let

(7) $\quad o = r \oplus r';$

then, by 9.1.4, also

(8) $\quad \boldsymbol{O}(opr').$

By the theorem 9.3.1, the formula (6) comined with (7) yields

(9) $\quad x_p = y_r \wedge y_p = -x_r \vee x_p = y_{r'} \wedge y_p = -x_{r'},$

which, combined with (5.1), gives $q = r$ or $q = r'$. Whence, by (6) and (8), we obtain $\boldsymbol{O}(opq)$. By the similar argument, the hypothesis

(5.2) $\quad x_p = -y_q \quad$ and $\quad y_p = x_q$

leads to the same conclusion. ∎

In view of 9.3.1 and 9.3.2 we have

Th E 9.3.3. $\boldsymbol{O}(opq) \Leftrightarrow x_p = y_q \wedge y_p = -x_q \vee x_p = -y_q \wedge y_p = x_q.$
Mid Des

We now pass to a basis with an arbitrary vertex a.

Th E 9.3.4. $\boldsymbol{O}(apq) \Leftrightarrow (x_p - x_a) = (y_q - y_a) \wedge (y_p - y_a) = -(x_q - x_a)$
Mid Des $\qquad \vee (x_p - x_a) = -(y_q - y_a) \wedge (y_p - y_a) = (x_q - x_a).$

P r o o f. Let

$$a \oplus o = p \oplus p' = q \oplus q';$$

then

(10) $\quad \Pi(p') = \Pi(a) - \Pi(p) \quad$ and $\quad \Pi(q') = \Pi(a) - \Pi(q).$

By 9.2.4

$$\boldsymbol{O}(apq) \Leftrightarrow \boldsymbol{O}(op'q'),$$

whence, by (10), in view of 9.3.3 we obtain the equivalence 9.3.4. ∎

Theorem 9.3.4 gives an analytic description of the relation $\boldsymbol{O}$.

Now, let us show that the scalar field $\mathfrak{F}$ is 2-formally real. Suppose that, on the contrary,

$$-1 = u^2$$

for some $u \in F$. Let

(11) $\Pi(p) = (1, u)$ and $\Pi(q) = (u, -1)$.

Then, by 9.3.4 we get

(12) $O(opq)$;

on the other hand, since $\Pi(q) = \Pi(p) \cdot u$, we have

(13) $L(opq)$.

However, by 9.1.11, the condition (12) combined with (13) implies $o = p$, which contradicts (11). Hence $1+u^2 \neq o$ for every $u \in F$.

Consider now the Cartesian structure

$$\underline{\mathfrak{A}}(\mathfrak{F}) = (F^2, \oplus_{\mathfrak{F}}, ||_{\mathfrak{F}})$$

and its expansion

$$\underline{\mathfrak{E}}(\mathfrak{F}) = (F^2, \oplus_{\mathfrak{F}}, ||_{\mathfrak{F}}, O_{\mathfrak{F}}).$$

By 6.4.3, the coordinate system Π maps isomorphically the midpoint plane $\mathfrak{A}$ onto the structure $\underline{\mathfrak{A}}(\mathfrak{F})$. In view of 9.3.4 and 9.1.12, this yields

Th E 9.3.5. $\mathfrak{E} \simeq \underline{\mathfrak{E}}(\mathfrak{F})$.
Mid Des

The function Π is an isomorphism.

Thus for every Desarguean Euclidean plane $\mathfrak{E}$ there is a 2-formally real field (namely the scalar field) $\mathfrak{F}$ such that $\mathfrak{E} \simeq \underline{\mathfrak{E}}(\mathfrak{F})$,

Theorem 9.3.5 combined with 9.1.13 results in

Representation Theorem 9.3.6. *Desarguean Euclidean planes coincide (up to isomorphism) with the Cartesian structures $\underline{\mathfrak{E}}(\mathfrak{F})$ over 2-formally real fields $\mathfrak{F}$.*

9.4. PAPPIAN EUCLIDEAN PLANES

A Desarguean Euclidean plane $\mathfrak{E} = (S, \oplus, ||, O)$ is said to be a *Pappian Euclidean plane* whenever it satisfies the axiom Papp. Then $\mathfrak{A} = (S, \oplus, ||)$ is a Pappian midpoint plane and therefore 6.4.4(iii) combined with 9.3.6 yields

Representation Theorem 9.4.1. *Pappian Euclidean planes coincide (up to isomorphism) with the Cartesian structures* $\mathfrak{E}(\mathfrak{F})$ *over 2-formally real commutative fields* $\mathfrak{F}$.

Take an arbitrary commutative and 2-formally real field $\mathfrak{F} = (F, 0, 1, +, \cdot)$. We are going to show that for every $x_1, y_1, x_2, y_2 \in F$ the following two conditions are equivalent:

(I) $\quad x_1 = y_2 \wedge y_1 = -x_2 \vee x_1 = -y_2 \wedge y_1 = x_2,$

(II) $\quad x_1 \cdot x_2 + y_1 \cdot y_2 = 0 \wedge x_1^2 + y_1^2 = x_2^2 + y_2^2.$

The inference (I) $\Rightarrow$ (II) is obvious. Assume conversely

(II.1) $\quad x_1 \cdot x_2 + y_1 \cdot y_2 = 0$

and

(II.2) $\quad x_1^2 + y_1^2 = x_2^2 + y_2^2.$

By (II.2), since $\mathfrak{F}$ is 2-formally real, we obtain the equivalence

$$(x_1, y_1) = (0, 0) \Leftrightarrow (x_2, y_2) = (0, 0).$$

If $(x_1, y_1) = (x_2, y_2) = (0, 0)$, then evidently the condition (I) holds. Assume $(x_1, y_1), (x_2, y_2) \neq (0, 0)$. Then, by (II.1)

$$x_1 = 0 \Leftrightarrow y_2 = 0.$$

If

$$x_1 = y_2 = 0,$$

then, in view of (II.2), by the commutativity of multiplication,

$$y_1 = -x_2 \quad \text{or} \quad y_1 = x_2$$

and therefore (I) holds again. Finally, if $x_1, y_2 \neq 0$, then (again by the commutativity of multiplication), by (II.1) we obtain

$$-x_2 \cdot y_2^{-1} = y_1 \cdot x_1^{-1}$$

and for some $u \in F$

(1) $\quad y_1 = u \cdot x_1 \quad$ and $\quad x_2 = -u \cdot y_2,$

whence, by (II.2),

$$(1+u^2) \cdot x_1^2 = (1+u^2) \cdot y_2^2.$$

Since $1+u^2 \neq 0$, we have $x_1^2 = y_2^2$ and consequently

$$x_1 = y_2 \quad \text{or} \quad x_1 = -y_2.$$

By (1), if $x_1 = y_2$, then also $y_1 = -x_2$ and if $x_1 = -y_2$, then $y_1 = x_2$. Thus again the condition (I) is satisfied. Therefore (II) $\models\!\Rightarrow$ (I). ∎

Let us now consider the Cartesian structure $\underline{\mathfrak{E}}(\mathfrak{F}) = (F^2, \oplus_{\mathfrak{F}}, \|_{\mathfrak{F}}, \boldsymbol{O}_{\mathfrak{F}})$. The relation $\boldsymbol{O}_{\mathfrak{F}}$ is defined by 9.1.12. Setting in the equivalence (I) $\rightleftarrows$ (II)

$$x_i = p_i - a_i \quad \text{and} \quad y_i = q_i - a_i \quad (i = 1, 2),$$

we obtain a new description of the relation $\boldsymbol{O}_{\mathfrak{F}}$:

Th E 9.4.2.[4] $\boldsymbol{O}_{\mathfrak{F}}(apq) \Leftrightarrow (p-a)\cdot(q-a) = 0 \wedge (p-a)^2 = (q-a)^2.$
Mid Papp

Consider a Pappian Euclidean plane $\mathfrak{E} = (\boldsymbol{S}, \oplus, \|, \boldsymbol{O})$ and choose parameters of coordinatization. Let $\mathfrak{F}$ be the scalar field and let Π be the rectangular coordinate system. Theorem 9.3.4 gives an analytic description of the relation $\boldsymbol{O}$ in the system Π. Setting in the equivalence (I) $\rightleftarrows$ (II)

$$x_1 = x_p - x_a, \quad y_1 = y_p - y_a, \quad x_2 = x_q - x_a, \quad y_2 = y_q - y_a,$$

we obtain a new analytic description of the relation $\boldsymbol{O}$:

Th E 9.4.3. $\boldsymbol{O}(apq) \Leftrightarrow (\Pi(p) - \Pi(a))\cdot(\Pi(q) - \Pi(a)) = o$
Mid papp $\wedge (\Pi(p) - \Pi(a))^2 = (\Pi(q) - \Pi(a))^2.$

Notice that in the presence of Papp the axiom 9.1.6 is superflous. Indeed, this axiom has been used only in the proof of 9.3.1; without use of 9.1.6 we have obtained

(i) $\quad \boldsymbol{O}(opq) \Rightarrow x_p^2 = y_q^2 \wedge y_p^2 = x_q^2;$

then, using 9.1.6 we have proved

(ii) $\quad \boldsymbol{O}(opq) \wedge (x_p = y_q \wedge y_p = x_q \vee x_p = -y_q \wedge y_p = -x_q)$

$$\Rightarrow x_p = o \wedge y_p = o.$$

Using the commutativity of multiplication, we derive (ii) from (i). Assume

(1) $\quad \boldsymbol{O}(opq);$

let

$$o = q \oplus q', \quad p \oplus q = o \oplus b, \quad p \oplus q' = o \oplus b';$$

---

[4] The term "Th E" indicates that the theorem concerns Pappian Euclidean
MidPapp
planes. (M.M.)

then

(2) $\Pi(b) = \Pi(p)+\Pi(q)$ and $\Pi(b') = \Pi(p)-\Pi(q)$

and, by 9.2.5,

(3) $O(abb')$.

By (3) combined with (i) we obtain

(4) $(x_p+x_q)^2 = (y_p-y_q)^2$ and $(y_p+y_q)^2 = (x_p-x_q)^2$.

Let

(5) $x_p = y_q \wedge y_p = x_q \vee x_p = -y_q \wedge y_p = -x_q$.

The formulae (4) and (5) yield

$$(x_p+y_p)^2 = (y_p-x_p)^2,$$

whence by the commutativity of multiplication we obtain $x_p = o$ or $y_p = o$. ∎

9.5. EQUIDISTANCE AND PERPENDICULARITY[5]

Let $\mathfrak{E} = (S, \oplus, ||, O)$ be a Pappian Euclidean plane and let $\mathfrak{F}$ be its scalar field (determined uniquely up to isomorphism—compare 4.5.7). In view of 9.4.1, the plane $\mathfrak{E}$ may be investigated either directly, by means of the axiomatic method, or indirectly, by means of the analytic method, i.e. by studying the Cartesian structure $\underline{\mathfrak{E}}(\mathfrak{F})$ isomorphic to $\mathfrak{E}$.

Perpendicularity and equidistance can be defined analytically as follows:

Def 9.5.1. $\perp(apq) \Leftrightarrow (\Pi(p)-\Pi(a)) \cdot (\Pi(q)-\Pi(a)) = o$.

Def 9.5.2. $pq \equiv rs \Leftrightarrow (\Pi(p)-\Pi(q))^2 = (\Pi(r)-\Pi(s))^2$.

From the definitions 9.5.1 and 9.5.2 the well-known basic properties of the relations $\perp$ and $\equiv$ can be easily derived. In particular it is easy to check that both relations are invariant under any reflection.

As a direct consequence of 9.4.3 we get

Th E 9.5.3. $O(apq) \Leftrightarrow \perp(apq) \wedge ap \equiv aq$.
Mid Papp

[5] M.M.—Based on the author's notes.

The relations $\perp$ and $\equiv$ can be also defined in terms of the primitive notions of the plane $\mathfrak{E}$:

Th E 9.5.4. $\perp(apq) \Leftrightarrow \exists r(\boldsymbol{O}(apr) \wedge \boldsymbol{L}(aqr))$.
Mid Papp

Th E 9.5.5. $ap \equiv ap' \Leftrightarrow \exists qq'(\boldsymbol{O}(apq) \wedge \boldsymbol{O}(ap'q') \wedge pp' \| qq')$.
Mid Mapp

We leave to the reader the simple proof of the theorem 9.5.4 (notice that we make an essential use of the commutativity of $\mathfrak{F}$); we pass to the proof of 9.5.5.

Assume first that $a = o$ and let

(0) $\Pi(p) = (x, y)$ and $\Pi(p') = (x', y')$.

Let $op \equiv op'$. Then, by the definition 9.5.2,

(1) $x^2 + y^2 = (x')^2 + (y')^2$.

We define the points q and q' by the conditions

(2) $\Pi(q) = (y, -x)$ and $\Pi(q') = (-y', x')$.

Then, by 9.3.2,

(3) $\boldsymbol{O}(opq)$ and $\boldsymbol{O}(op'q')$.

Let us show that

(4) $pp' \| qq'$.

Indeed,

(5) $\Pi(p) - \Pi(p') = (x - x', y - y')$

and

(6) $\Pi(q) - \Pi(q') = (y + y', -(x + x'))$,

and the proportionality of these two pairs of scalars can be expressed (in a commutative field) by the condition

$$-(x - x') \cdot (x + x') = (y - y') \cdot (y + y'),$$

which is equivalent to (1).

Conversely, if the points q and q' satisfy the conditions (3) and (4), then, by (0), in view of 9.3.1

$$\Pi(q) = (y, -x) \quad \text{or} \quad \Pi(q) = (-y, x)$$

and

$$\Pi(q') = (y', -x') \quad \text{or} \quad \Pi(q') = (-y', x'),$$

whence the point $\Pi(q)-\Pi(q')$ coincides with one of the following four points of the set F^2:

$$(y-y', -(x-x')), \quad (-(y-y'), x-x'),$$
$$(y+y', -(x+x')), \quad (-(y+y'), x+x').$$

Thus, by (2) and (4),

$$-(x-x')\cdot(x-x') = (y-y')\cdot(y-y')$$

or

$$-(x-x')\cdot(x+x') = (y-y')\cdot(y+y'),$$

i.e.

$$(x-x')^2+(y-y')^2 = o \quad \text{or} \quad x^2+y^2 = (x')^2+(y')^2.$$

However, since $\mathfrak{F}$ is 2-formally real, the first of the last two equalities implies $p = p'$, whence, by 9.5.2, in both cases $op \equiv op'$.

The general case can be reduced to the particular case ($a = o$) by means of the reflection with respect to the point $a \oplus o$; indeed, by 9.2.4 and 6.3.1 this reflection preserves the relations $\boldsymbol{O}$ and $\|$, and, by the definition 9.5.2, also the relation $\equiv$. ■

As a direct consequence of 9.5.2 and 6.3.8, we obtain

Th 9.5.6. $\quad ap \equiv bq \Leftrightarrow \exists p'\; ap \equiv ap' \wedge a\oplus q = b\oplus p'.$
Mid

Theorem 9.5.6 together with 9.5.5 describes the relation $\equiv$ in terms of the primitive notions $\oplus$, $\|$, and $\boldsymbol{O}$ of the plane $\mathfrak{E}$.

BIBLIOGRAPHY

Bachmann, F.

[59] *Aufbau der Geometrie aus dem Spiegelungsbegriff*, Springer, Berlin, 1959.

Doraczyńska, E.

[74] *A complete theory of the midpoint operation*, Bull. Acad. Polon. Sci. Sér. Sci. Math., Astronom., Phys. 22, 1974, pp. 1195–1200.

Hauschild, K. und Rautenberg W.

[74] *Rekursive Unentscheidbarkeit der Theorie der pythagoräischen Körper*, Fund. Math. 82, 1974, pp. 191–197.

Hilbert, D.

[50] *Foundations of Geometry*, The Open Court Publishing Company, La Salle, Illinois, 1950.

Kordos, M.

[69] *On the syntactic form of dimension axiom for affine geometry*, Bull. Acad. Polon. Sci. Sér. Sci. Math., Astronom., Phys. 17, 1969, pp. 833–837.

Kurosh, A.G.

[73] *Lectures on Universal Algebra* (in Russian), Moskva, 1973.

Lenz, H.

[61] *Grundlagen der Elementarmathematik*, VEB Deutscher Verlag der Wissenschaften, Berlin, 1961.

Lingenberg, R. and Baur, A.

[60] *Der synthetische und der analytische Standpunkt in der Geometrie*, Grundzüge der Mathematik, Band II Geometrie, Vandenhoeck & Ruprecht, Göttingen, 1960.

Loś, J. and Suszko R.

[57] *On the extending of models (IV)*, Fund. Math. 44, 1957, pp. 52–60.

Moszyńska, M.

[77] *Theory of equidistance and betweenness relations in regular metric spaces*, Fund. Math. 96, 1977, pp. 17–29.

Pickert, G.

[55] *Projektive Ebenen*, Springer, Berlin, 1955.

[75] *Projektive Ebenen*, Springer, Berlin, 1975.

Piesyk, Z.

[77] *On the betweenness relation in a set of cardinality* $\neq 4$, Bull. Acad. Polon. Sci. Sér. Sci. Math., Astronom., Phys. 25, 1977, pp. 667–670.

Szczerba, L.

[70] *Independence of Pasch's axiom*, Bull. Acad. Polon. Sci. Sér. Sci. Math., Astronom., Phys. 18, 1970, pp. 491–498.

[71a] *Incompleteness degree of elementary Pasch-free geometry*, Bull. Acad. Polon. Sci. Sér. Sci. Math., Astronom., Phys. 19, 1971, pp. 213–215.

[71b] *Undefinability of order in Pasch-free geometry*, Ibid. pp. 315–317.

[71c] *Undecidability of elementary Pasch-free geometry*, Ibid. pp. 469–474.

[71d] *The Pasch-free geometry with the continuity axiom*, Ibid. pp. 313–616.

Szczerba, L. and Szmielew, W.

[70] *On the Euclidean geometry without the Pasch axiom*, Bull. Acad. Pol. Sci. Sér. Sci. Math., Astronom., Phys. 18, 1970, pp. 659–666.

Szczerba, L. and Tarski, A.

[79] *Metamathematical discussion of some affine geometries*, Fund. Math. 104, 1979, pp. 155–192.

Szmielew, W.

[70] *The Pasch axiom as a consequence of the circle axiom*, Bull. Acad. Polon. Sci. Sér. Sci. Math., Astronom., Phys. 18, 1970, pp. 751–758.

[74a] *The order and the semi-order of n-dimensional Euclidean space in the axiomatic and model-theoretic aspects*, Grundlagen der Geometrie und algebraische Methoden, Potsdamer Forschungen—Reihe B, Heft 3, Potsdam, 1974, pp. 69–70.

[74b] *The role of the Pasch axiom in the foundations of Euclidean geometry*, Proc. Tarski Symp., 1974, pp. 123–132.

[77] *Oriented and non-oriented linear orders*, Bull. Acad. Polon. Sci. Sér. Sci. Math., Astronom., Phys. 25, 1977, pp. 659–665.

[79] *Concerning the order and the semi-order of n-dimensional Euclidean space*, Fund. Math. 107, 1979, pp. 47–56.

[80] *Theory of n-ary equivalences and its application to geometry*, Diss. Math. 191, 1980.

Tarski, A.

[59] *What is elementary geometry*? Studies in Logic and the Foundations of Mathematics, Proc. Internat. Sympos. North-Holland, Amsterdam, 1959, pp. 16–29.

Van der Waerden, B.L.

[53] *Modern Algebra*, vol. 1, New York, 1953.

Włodarski, L.

[79] *On the Pappus, Desargues and related theorems of affine geometry*, Bull. Acad. Polon. Sci. Sér. Sci. Math., Astronom., Phys. 27, 19 p. 869–873.

BIBLIOGRAPHY OF THE WORKS OF WANDA SZMIELEW

[46] *On choices from finite sets*, Fund. Math. 34, 1946, pp. 75–80.

[49a] *Decision problem in group theory*, Proceedings of the X-th International Congress of Philosophy, Amsterdam 1949, pp. 763–766.

[49b] *Arithmetical classes and types of Abelian groups*, Bull. Amer. Math. Soc. 55, 1949, pp. 65.

[49c] (and A. Tarski) *Theorems common to all complete and axiomatizable theories*, ibidem 55, 1949, pp. 1075.

[50] (and A. Tarski) *Mutual interpretability of some essentially undecidable theories*, Proceedings of the International Congress of Mathematicians, Cambridge 1950, pp. 734.

[54] *Elementary properties of Abelian groups*, Fund. Math. 41, 1954, pp. 203–271.

[59a] *Some metamathematical problems concerning elementary hyperbolic geometry*, The Axiomatic Method, Amsterdam 1959, pp. 30–52.

[59b] *Absolute calculus of segments and its metamathematical implications*, Bull. Acad. Polon. Sci. Sér. Sci. Math., Astronom., Phys. 7, 1959, pp. 213–220.

[60] (and K. Borsuk) *Foundations of geometry*, Studies in Logic and the Foundations of Mathematics, Amsterdam 1960, p. 444 (also in Polish).

[61] *A new analitic approach to hyperbolic geometry*, Fund. Math. 50, 1961, pp. 129–158.

[62] *New foundations of absolute geometry*, Methodology and Philosophy of Science, Standford University Press, 1962, pp. 168–175.

[70a] (and L. W. Szczerba) *On the Euclidean geometry without the Pasch axiom*, Bull. Acad. Polon. Sci. Sér. Sci. Math., Astronom., Phys. 18, 1970, pp. 659–666.

(70b] *The Pasch axiom as a consequence of the circle axiom*, ibidem 18, 1970. pp. 751–758.

[70c] *A statement on two circles as the geometric analog of Euclid's field property*, ibidem 18, 1970, pp. 759–764.

[74a] *The order and the semi-order of n-dimensional Euclidean space in the axiomatic and model-theoretic aspects*, Grundlagen der Geometrie und algebraische Methoden, Potsdamer Forschungen—Reihe B, Heft 3, 1974, pp. 69–80.

[74b] *The role of the Pasch axiom in the foundations of Euclidean geometry*, Proc. Tarski Symp. 1974, pp. 123–132.

[77] *Oriented and nonoriented linear orders*, Bull. Acad. Polon. Sci. Sér. Sci. Math. Astronom., Phys. 25, 1977, pp. 659–665.

[79] *Concerning the order and the semi-order of n-dimensional Euclidean space*, Fund. Math. 107, 1979, pp. 47–56.

[80] *Theory of n-ary equivalences and its application to geometry*, Diss. Math, 191, 1980.

[81] *Od geometrii afinicznej do Euklidesowej — Rozważania nad aksjomatyką*, PWN, Warszawa, 1981 (English translation — see [83]).

[83] *From affine to Euclidean geometry. An axiomatic approach*, D. Reidel Publishing Company, Dordrecht, Boston, London and PWN, Warsaw, 1983.

LIST OF ALGEBRAIC SYMBOLS

0	zero of a loop $(G, 0, +)$
1	unit of a loop $(G, 1, \cdot)$
$+$	addition
$\cdot$	multiplication
$-a$	element opposite to a
a^{-1}	— inverse to a
$\oplus$	midpoint operation
$\leqslant$	ordering relation
$\mathfrak{F}$	quasi-field (field)
$\mathfrak{T}$	ternary field
$\mathfrak{F}_+$	additive group of a quasi-field $\mathfrak{F}$
$\mathfrak{F}.$	multiplicative group of a field $\mathfrak{F}$
$\mathfrak{G}^2$	Cartesian square of a group $\mathfrak{G}$
$\underline{\mathfrak{T}}(\mathfrak{F})$	ternary field of a quasi-field $\mathfrak{F}$
$\underline{\mathfrak{V}}^2(\mathfrak{F})$	right Cartesian (2-dimensional) linear space over $\mathfrak{F}$
$\mathbf{TF}$	class of ternary fields
$\mathbf{QF}$	— — strong left quasi-fields
$\mathbf{F}$	— — fields
com $\mathbf{F}$	— — commutative fields
$\mathbf{OQF}$	— — ordered strong left quasi-fields
$\mathbf{OF}$	— — — fields
com $\mathbf{OF}$	— — — commutative fields
$\mathbf{G}_{\left(\frac{1}{2}\right)}$	— — groups with the operator $\frac{1}{2}$
$\mathbf{M}_{(\mathrm{zero})}$	— — midpoint algebras with zero

LIST OF GEOMETRIC SYMBOLS

$a, b, c, d, e, o, p, q, r, s$	points
K, L, M, P, Q, R	lines
$\boldsymbol{S}$	universe
$\boldsymbol{B}$	betweenness relation
$\boldsymbol{L}$	collinearity —
$\parallel$	parallelity —
$\boldsymbol{P}$	parallelogram —
$\boldsymbol{T}_0$	trapezoid —
$\boldsymbol{P}_M$	extended parallelogram —
$\boldsymbol{O}$	equi-orthogonal basis —
$\perp$	perpendicularity —
$\triangle$	Pieri's —
$\equiv$	equidistance —
$[ab]$	set of points collinear with a, b
$L_{\parallel P}(p)$	line parallel to P and passing through p
$\pi\colon K \underset{\parallel P}{\rightarrow} L$	parallel projection of K onto L in direction of P
δ	dilatations
α, β, γ	translations
$\varkappa\ \lambda$	homotheties
$\tau(\mathrm{ab})$	translation carrying a onto b
$\iota_0(\mathrm{ab})$	homothety with center o carrying a onto b
$+$	composition of translations
$\cdot$	— — homotheties
$+_o$	addition dependent on a parameter o
$\cdot_o^a$	multiplication dependent on parameters o and a
$(o, 1, e)$	parameters of coordinatization
Aut	group of automorphisms of parallelity plane
Dil	— — dilatations
Ht_o	— — homotheties with center o
Ht	set of homotheties
Tr	group of translations
Tr_P	— — — in direction of P
$\underline{\mathfrak{A}}(\mathfrak{T})$	parallelity plane over a ternary field $\mathfrak{T}$
$\underline{\mathfrak{A}}(\mathfrak{F})$	— — — a quasi-field $\mathfrak{F}$
$\underline{\mathfrak{E}}(\mathfrak{F})$	Euclidean plane over a field $\mathfrak{F}$
$\underline{\mathfrak{A}}(\mathfrak{F}, \leqslant)$	Cartesian ordered midpoint plane over an ordered field $(\mathfrak{F}, \leqslant)$
$\mathscr{A}$	class of parallelity planes

des $\mathscr{A}$	class of translation planes
Des $\mathscr{A}$	— — homothety planes
Papp $\mathscr{A}$	— — commutative homothety planes
Mid $\mathscr{A}$	— — midpoint planes
Mid Des $\mathscr{A}$	— — Desarguean midpoint planes
Mid Papp $\mathscr{A}$	— — Pappian midpoint planes
OMid Des $\mathscr{A}$	— — ordered Desarguean midpoint planes
OMid Papp $\mathscr{A}$	— — — Pappian midpoint planes
$\|_{\mathfrak{F}}, L_{\mathfrak{F}}, B_{\mathfrak{F}}, O_{\mathfrak{F}}$	parallelity, collinearity, ... relations on the plane over a field $\mathfrak{F}$
$\|_{\mathfrak{T}}, L_{\mathfrak{T}}$	parallelity and collinearity relations on the plane over a ternary field $\mathfrak{T}$
$\Pi(p)$	coordinate system of a point p
$p(x, y)$	point with coordinates x, y

INDEX

addition 1
additive group of a line 54
affine plane 16
Archimedean order in a group 129
— — — — midpoint algebra 140
arithmetic mean 89
axis 71

basis 162
betweenness relation 110
— — compatible with midpoint operation 145
— — on a plane 146

Cartesian coordinate system 73
— ordered midpoint plane 158
— square of a group 10
characteristic of a translation plane 96
collinearity 12
— structure 12
commutative field 6
— group 3
— homothety plane 52
— multiplicative group with zero 4
continuity axiom 117
continuous betweenness relation 117
— order 116
cyclic element of a group 129

Dedekind section 117
density of order 120
Desarguean affine plane 43
— Euclidean plane 165
diagonal of a parallelogram 27
dilatation 30
dimension of a collinearity structure 13
direction of a translation 32
double loop 4

equidistance 162
equi-orthogonal basis 162

first coordinate 73
field 5
— of a line 64

geometric midpoint 100
group 2
— with the operator $\frac{1}{2}$ 89
groupoid 1

half 88
homothety plane 38
— with center *o* 30

improper center of a translation 32
inverse element in a group 3

left quasi-field 5
line 14
linear order 110
— parallelogram 105
linearly ordered set 110
loop 1

major Desargues law 40
midpoint of a pair 89
— algebra 89
— — with zero 90
— operation 89
— plane 104
minor Desargues law 40
multiplication 3
— by **2** 90
multiplicative group with zero 4
— — of a line 60
— loop with zero 3

non-oriented order 127

operator $\frac{1}{2}$ 89
— **2** 91
opposite element in a group 3
order 110

ordered Desarguean midpoint plane 156
— group 126, 127
— left quasi-field 132
— midpoint algebra 138
— — plane 155
— Pappian midpoint plane 156
— translation plane 153
ordering of a quasi-field 132

Pappian affine plane 52
— Euclidean plane 180
parallel projection 24
parallelity plane 19
parallelogram 27
— relation 26
partial order in a group 130
perpendicularity 162
plane parallelogram 105
positive cone 128
pseudo-axis 71
pseudo-scalar 71

reflexivity of a ternary relation 12
right quasi-field 5

second coordinate 73
segment 17
scalar 71
side of a parallelogram 27
strong left quasi-field 8
strongly Desarguean affine plane 43
subtraction 3
symmetry of a ternary relation 12

ternary field 6
transitive group 38
transitive set 38
transitivity of a ternary relation 12
translation 30
— plane 38
trapezoid 29
— relation 29
two-formally real field 167
two-sided quasi-field 5

unit of a loop 3

weakly Desarguean affine plane 43

zero of a loop 1

GPSR Compliance

The European Union's (EU) General Product Safety Regulation (GPSR) is a set of rules that requires consumer products to be safe and our obligations to ensure this.

If you have any concerns about our products, you can contact us on ProductSafety@springernature.com

In case Publisher is established outside the EU, the EU authorized representative is:

Springer Nature Customer Service Center GmbH
Europaplatz 3
69115 Heidelberg, Germany

Batch number: 10584448

Printed by Printforce, the Netherlands